Cucurbit Genetics Cooperative

33-34

2010-2011

ISSN 1064-5594

2102 Plant Sciences Building
College Park, Maryland
20742-4452 USA

Tel: (301) 405-1321 or (580) 889-7395
Fax: (301) 314-9308

cucurbit.genetics.cooperative@gmail.com
angela.davis@lane-ag.org

http://cuke.hort.ncsu.edu/cgc/
http://www.umresearch.umd.edu/cgc/

CGC Coordinati[ng]

Chair:

Associate Chairs:

Print Edition	Angela R. Davis Lane, OK, USA
Membership	Linda Wessel-Beaver Mayagüez, PR, USA
Assistant	Zahyong Sun Nanning, Guangxi, China
Treasurer	Timothy J Ng College Park, MD, USA

Assistant Editors:

Cucumber:	Rebecca Grumet East Lansing, MI , USA
Melon:	Kevin Crosby College Station, TX, USA
Watermelon:	Amnon Levi Charleston, SC, USA
2010	Stephen King College Station, TX USA
Cucurbita spp.:	James Myers Corvallis, OR, USA
Other genera: Current	Tusar Behera Indian Ag. Res. Inst, India

CGC Gene Curators and Gene List Committee

Cucumber:	Yiqun Weng Madison, WI, USA Todd C. Wehner Raleigh, NC, USA
Melon:	Catherine Dogimont Montfavet, France Michel Pitrat Montfavet, France James D. McCreight Salinas, CA, USA
Watermelon:	Todd C. Wehner Raleigh, NC, USA Stephen R. King College Station, TX, USA
Cucurbita spp.:	Harry Paris Ramat Yishay, Israel Les Padley Naples, FL, USA
Other genera:	Mark G. Hutton Monmouth, ME, USA Thomas C. Andres Bronx, NY, USA

The **Cucurbit Genetics Cooperative** (CGC) was organized in 1977 to develop and advance the genetics of economically important cucurbits. Membership to CGC is voluntary and open to individuals who have an interest in cucurbit genetics and breeding. CGC membership is on a biennial basis. For more information on

CGC and its membership rates, visit our website (*http://cuke.hort.ncsu.edu/cgc*)
or contact Tim Ng, (301)405-1321, *cucurbit.genetics.cooperative@gmail.com*, or Angela Davis, (580)889-7395,
angela.davis@lane-ag.org).

ISSN 1064-5594

Contents

Other Genera and Species

GENE LISTS

MEMBER DIRECTORIES

News & Comments

34th CGC Business Meeting (2010)

Todd C. Wehner, Department of Horticultural Science, North Carolina State University, Raleigh, NC 27695-7609 USA

The Cucurbit Genetics Cooperative met with the Cucurbitaceae 2010 conference in Charleston, South Carolina at 6:30 pm on November 15. Editors were elected to replace those who were retiring.

Linda Wessel-Beaver presented an overview of CGC membership. The membership list has been updated. Conference attendees were permitted to register for CGC membership on-site.

Angela Davis presented summary statistics and cost analyses on the annual CGC Reports. A paper edition will continue to be made available, this time in a smaller format.

The CGC website has been updated and expanded. All back issues have been put onto the world wide web, along with all issues of the Vegetable Improvement Newsletter. Help is needed to review past CGC articles and report any corrections needed or typographical errors found.

Announcements were made on the upcoming EU-CARPIA Cucurbit meeting in Antalya, Turkey in 2012, as well as plans for Cucurbitaceae 2014.

Comments from the CGC Coordinating Committee

The Call for Papers for the 2011 Report (CGC Report No. 34) has been sent out. Papers should be submitted to the respective Coordinating Committee members for publication in the volume. As always, we are eager to hear from CGC members regarding our current activities and the future direction of CGC.

- Todd C. Wehner, chair and website editor
- Angela Davis, associate chair and print editor
- Linda Wessel-Beaver, associate chair and membership coordinator
- Tim Ng, associate chair and treasurer
- Rebecca Grumet, assistant editor (cucumber)
- Kevin Crosby, assistant editor (melon)
- Tusar Behera, assistant editor (other genera)
- James Myers, assistant editor (*Cucurbita* spp.)
- Stephen R. King, assistant editor (watermelon)

Cucurbit Genetics Cooperative Report Call for Papers

The call for papers for **CGC 35** (2012) is open, and we are **accepting papers** for the volume now. Send manuscripts to the appropriate crop editor. (*http://cuke. hort.ncsu.edu/cgc*) If you do not receive your copy, contact Linda Wessel-Beaver: lindawessel.beaver @upr.edu.

Comments from CGC Gene List Committee

Lists of known genes for the Cucurbitaceae have been published previously in Hortscience and in reports of the Cucurbit Genetics Cooperative. CGC is currently publishing complete lists of known genes for cucumber (*Cucumis sativus*), melon (*Cucumis melo*), watermelon (*Citrullus lanatus*), *Cucurbita* spp., and other cucurbit genera and species on a rotating basis.

It is hoped that scientists will consult these lists as well as the rules of gene nomenclature for the Cucurbitaceae before choosing a gene name and symbol. Thus, inadvertent duplication of gene names and symbols will be prevented. The rules of gene nomenclature were adopted in order to provide guidelines for the naming and symbolizing of genes previously reported and those which will be reported in the future. Scientists are urged to contact members of the Gene List Committee regarding questions in interpreting the nomenclature rules and in naming and symbolizing new genes.

- Cucumber: Yiqun Weng (curator) and Todd C. Wehner (assistant)
- Melon: Catherine Dogimont (curator) Michael Pitrat (assistant curator) and James D. McCreight (assistant curator)
- Other Genera: Mark G. Hutton (curator) and Thomas Andres (assistant curator)
- *Cucurbita* spp.: Harry Paris (curator) and Les Padley (assistant curator)
- Watermelon: Todd C. Wehner (curator) and Stephen R. King (assistant curator)

Comments from CGC Gene Curators

Lists of known genes for the Cucurbitaceae have been published in reports of the Cucurbit Genetics Cooperative. CGC is currently publishing complete lists of known genes for cucumber (*Cucumis sativus*), melon (*Cucumis melo*), watermelon (*Citrullus lanatus*), *Cucurbita* spp., and other genera on a 5-year rotation.

We hope that scientists will consult these lists as well as the rules of gene nomenclature for the Cucurbitaceae before choosing a gene name and symbol. In this way, we hope to avoid inadvertent duplication of gene names and symbols. The rules of gene nomenclature were adopted in order to provide guidelines for the naming and symbolizing of genes previously reported and those which will be reported in the future. Scientists are urged to contact members of the Gene List Committee regarding questions in interpreting the nomenclature rules and in naming and symbolizing new genes.

CGC has appointed Curators for the major cultivated groups: cucumber, melon, other genera, *Cucurbita* spp, and watermelon.

Curators are responsible for collecting, maintaining and distributing upon request stocks of the known marker genes. CGC members are requested to forward samples of currently held gene stocks to the respective Curator.

- Cucumber: Yiqun Weng (curator) and Todd C. Wehner (assistant curator)
- Melon: Catherine Dogimont (curator) and James D. McCreight (assistant curator)
- Other Genera: Mark G. Hutton (curator) and Thomas Andres (assistant curator)
- *Cucurbita* spp.: Harry Paris (curator) and Richard W. Robinson (assistant curator)
- Watermelon: Todd C. Wehner (curator) and Stephen R. King (assistant curator)

2010 Watermelon Research and Development Working Group – 30th Annual Meeting

Submitted by Jonathan R. Schultheis

The Annual Meeting of the Watermelon Research & Development Working Group was held Sunday, February 7, 2010 at the Wyndham Orlando Resort in Orlando, FL, from 8:00 a.m. to 4:00 p.m. The meeting was held in conjunction with The Southern Association of Agricultural Scientists and the Southern Region American Society for Horticultural Science (SR-ASHS). A welcome from Elisabetta Vivoda, chair of the Watermelon Research & Development Working Group (WRDWG), was given to all in attendance which totaled near 30 people. Judy Thies, chair of the upcoming international Cucurbitaceae meeting, reminded and invited everyone at the WRDWG meeting that Cucurbitaceae 2010 would be held in Charleston, SC, November 14-18, 2010.

An update in watermelon cultivar development was then provided from the following seed companies: Harris Moran (Brenda Lanini), Syngenta (James Brusca), and Zeraim Gedera (Woody Spiers).

Watermelon trial reports were then given by Jonathan Schultheis: Red flesh watermelon and mini-watermelon trials, 2009; and Brad Thompson: Yellow orange flesh watermelon trial, 2008. After their presentations, there was a lengthy discussion regarding watermelon quality attributes with particular interest in hollow heart.

After the watermelon cultivar trial results the following research reports were presented the remainder of the morning and after lunch:

Lagenaria and *Cucurbita* **rootstocks prevent infection of watermelon scions by** *Fusarium oxysporum* **f. sp.** *niveum.* **A. P. Keinath***, V. B. DuBose, and R. L. Hassell. Clemson University, Coastal REC, Charleston, SC.

Tolerance to the herbicide clomazone in watermelon plant introductions. H. Harrison, **C.S. Kousik** *and A. Levi. U.S. Vegetable Laboratory, USDA ARS, Charleston, SC.

Phylogenetic Relationships Among Cucurbit Species Used as Rootstocks for Grafting Watermelon. A. Levi*[1], J.A. Thies[1], K. Ling[1], A. Simmons[1], C.S. Kousik[1], W.P. Wechter[1], and R. Hassell[2]. [1]USDA-ARS, U.S. Vegetable Laboratory, and [2]Clemson University, Costal Research and Education Center, 2700 Savannah Highway, Charleston, SC 29414, USA 2700 Savannah Highway, Charleston, SC.

Isolation, Sequence Analysis, and Linkage Mapping of Disease Resistance Gene Analogs in Watermelon. K. R. Harris, **W. P. Wechter***, and A. Levi USDA-ARS, U.S. Vegetable Laboratory, 2700 Savannah Highway, Charleston, SC.

Update on the watermelon vine decline virus and other whitefly-transmitted cucurbit viruses in Florida, and their effects on watermelon. Scott Adkins*[1], Craig G. Webster[1], T. Greg McCollum[1], Joseph P. Albano[1], Chandrasekar S. Kousik[2], Pamela D. Roberts[3], Susan E. Webb[4], Carlye A. Baker[5] and William W. Turechek[1]. [1]USDA, ARS, Fort Pierce, FL, [2]USDA, ARS, Charleston, SC, [3]SWFREC, University of Florida, Immokalee, FL, [4]University of Florida, Gainesville, FL, [5]FDACS-DPI, Gainesville, FL.

Effects of the watermelon vine decline virus on wild watermelon germplasm and other vining cucurbits.. Craig G. Webster[1]*, Chandrasekar S. Kousik[2], William W. Turechek[1], Pamela D. Roberts[3], Susan E. Webb[4] and Scott Adkins[1]. [1]USDA-ARS, U.S. Horticultural Research Laboratory, Fort Pierce, FL, [2]USDA-ARS, U.S. Vegetable Laboratory, Charleston, SC, [3]SWFREC, University of Florida, Immokalee, FL, [4]University of Florida, Gainesville, FL.

Patterns of Virus Distribution in Single and Mixed Infections of Florida Watermelons. William W. Turechek, U.S. Department of Agriculture, Agricultural Research Service, U.S. Horticultural Research Laboratory, 2001 South Rock Road, Fort Pierce, FL 34945; **Chandrasekar S. Kousik,** U.S. Department of Agriculture, Agricultural Research Service, U.S. Vegetable Laboratory, 2700 Savannah Highway, Charleston, SC 29414; and **Scott Adkins**, U.S. Department of Agriculture, Agricultural Research Service, U.S. Horticultural Research Laboratory, 2001 South Rock Road, Fort Pierce, FL.

Resistance of wild watermelon (*Citrullus lanatus* var. *citroides*) rootstocks to southern root-knot nematode. **Judy A. Thies*** (1), Jennifer J. Ariss (1), Amnon Levi (1), Chandrasekar S. Kousik (1), and Richard L. Hassell (2). (1) U.S. Vegetable Laboratory, USDA, ARS, Charleston, SC; (2) Clemson Coastal Research & Education Center, Charleston, SC.

Evaluation of instrumental texture measurements of watermelon cultivars. Jennifer W. Shiu*, University of California Davis, Davis, CA; David C. Slaughter, University of California Davis, Davis, CA; Laurie Boyden, Syngenta Seeds, Naples, FL; Diane M. Barrett, University of California Davis, Davis, CA.

*Denotes who made presentation.

Following the more formal presentations, Shaker Kousik led a good discussion session aimed at answering some of the following questions: How many watermelon varieties are grown in the US? Can the name Hollow Heart be changed to Sweet heart since most watermelons with hollow heart are generally sweeter?

Elisabetta Vivoda presided over the WRDWG meeting as this was the first year of a two year term as chair of the group. She did a wonderful job in organizing the meeting.

Jonathan Schultheis served the first year of a two year term as vice-chair.

Todd Wehner was re-elected secretary.

The WRDWG thanks Syngenta for providing refreshments at the meeting.

2011 Watermelon Research and Development Working Group – 31ˢᵗ Annual Meeting

Submitted by Jonathan R. Schultheis and Richard Hassell

Prior to the 31ˢᵗ Annual Meeting of the Watermelon Research & Development Working Group (WRDWG), a session was held for the WRDWG at the Cucurbitaceae 2010 meeting held in Charleston, South Carolina in November 2010. About 60 people attended the WRDWG session at Cucurbitaceae 2010. One of the focuses of the session was to gain perspectives on watermelon production practices and challenges across various international geographic regions. This provided the opportunity for the group to gain some international perspectives rather than only a national perspective. Several watermelon breeders familiar with various production regions within Europe and Asia willingly shared their knowledge with the group. Additionally, there was discussion as to whether to have the WRDWG meeting in Corpus Christi. Although many who attended Cucurbitaceae could not or chose not to attend the meeting in Corpus Christi, there was a critical mass with several excellent presentations such that the 31ˢᵗ meeting did take place.

The Annual Meeting of the Watermelon Research & Development Working Group (WRDWG) was held Sunday, February 6, 2010 at Corpus Christi, Texas, from 8:00 a.m. to 12:00 noon. The meeting was held in conjunction with The Southern Association of Agricultural Scientists and the Southern Region American Society for Horticultural Sciences (SR-ASHS). A welcome from Richard Hassell, substituting as chair of the WRDWG, was given to all in attendance which totaled near 25 people. The weather, in addition to the close proximity of the recent cucurbit meeting in Charleston, South Carolina resulted in a lower attendance.

An update in watermelon cultivar development was then provided from the following seed companies: Syngenta and Zeraim Gedera. The following reports were given:

Irrigation Levels Affect Fruit Yield and Quality of Watermelon. Juan C. Díaz-Pérez, Dan MacLean, Pingsheng, Department of Horticulture, Tifton Campus, University of Georgia 31793 (jcdiaz@uga.edu)

Tolerance to Phytophthora Fruit Rot in Watermelon Plant Introductions. C.S. Kousik, U.S. Vegetable Laboratory, USDA, ARS, 2700 Savannah Highway, Charleston, SC 29414 (shaker.kousik@ars.usda.gov)

Results of 2010 Fungicide Trials to Manage Phytophthora Fruit Rot of Watermelon in South Carolina. C.S. Kousik, H.F. Harrison and J.A. Thies, U.S. Vegetable Laboratory, USDA, ARS, 2700 Savannah Highway, Charleston, SC 29414 (shaker.kousik @ars.usda.gov)

Performance of grafted watermelon in root-knot nematode infested soils. Judy A. Thies, U.S. Vegetable Laboratory, USDA, ARS, Charleston, SC, 29414 (Judy. Thies@ars.usda.gov) Richard L. Hassell, Clemson University CREC, 2700 Savannah Highway, Charleston, SC 29414 Jennifer J. Ariss, and Amnon Levi, U.S. Vegetable Laboratory, USDA, ARS, Charleston, SC 29414

Field Survey of Pollenizer Flowering, Triploid Fruit Set, and Pollinator Activity in Delaware Watermelons. Gordon C. Johnson, University of Delaware Carvel REC, 16483 County Seat Highway, Georgetown, DE 19947 (gcjohn@udel.edu)

QTL Mapping of Important Horticultural Traits in Watermelon. Cecilia E. McGregor. Department of Horticulture, University of Georgia, Athens, GA 30602 (cmcgre1@uga.edu)

Watermelon Fruit Quality Study 2010. Richard L. Hassell, Clemson University CREC, 2700 Savannah Highway, Charleston, SC. 29414 (rhassel@clemson.edu),

Penelope Perkins-Veazie, NC Research Campus, Kannapolis NC 28083 penelope_perkins@ncsu.edu

Comment from the U.S. Cucurbit Crop Germplasm Committee Chair

J.D. McCreight, USDA-ARS, Salinas, California USA.

The Cucurbit Crop Germplasm Committee (CCGC) operates under the auspices of the USDA-ARS National Plant Germplasm System (NPGS), is composed of ARS, university and industry scientists, and provides guidance to NPGS on matters relating to cucurbit crop and wild related species. CCGC membership and species-specific crop reports are accessible through the NPGS website: (http://www.ars-grin.gov/npgs/). The CCGC receives, reviews, and recommends germplasm evaluation proposals annually for funding by NPGS, and also reviews and recommends proposals for germplasm collections and exchange. Contact James D. McCreight (james.mccreight@ars.usda.gov) for more information. In 2010, the CCGC met with the Cucurbitaceae 2010 conference in Charleston, South Carolina at 6:00 P.M. on November 16. Reports were provided by NPGS on recent germplasm exploration activities and imminent release of GRIN Global software. Curator reports were presented. Proposed descriptors for Buffalo gourd (*Cucurbita foetidissima*) were submitted by the curator for review by the CCGC.

Upcoming Meetings of Interest to Cucurbit Researchers

Cucurbitaceae 2012

October 15-18, 2012, Antalya, Turkey

Dear Colleagues,

On behalf of the EUCARPIA Vegetable section, the conference organizing committee is pleased to welcome you to Antalya, Turkey, for the Xth EUCARPIA Meeting on Genetics and Breeding of Cucurbitaceae 2012.

The meeting will take place in Antalya. The city has always been a popular destination due to charming geography, history, and culture.

The EUCARPIA meetings on Cucurbitaceae are held every 4 years. Cucurbitaceae 2012 aims to bring together all researchers working on cucurbit genetics and breeding.

Cucurbitaceae 2012 will focus on all aspects of:
* Genetic resources
* Genetics and breeding
* Genomics and biotechnology

The scientific program includes oral and poster presentations submitted to the scientific committee. The language of the meeting will be English. The meeting will include a one-day technical excursion. Submission of posters and oral communications will be accepted August 01, 2011 - March 15, 2012. Online registration opens August 01, 2011. Deadline for reduced registration fee is April 15, 2012. Final registration (regular fee) August 15, 2012.

With Warm Regards,
The Organizing Committee and the Organizing President Prof. Dr. Nebahat SARI;
____nesari@cu.edu.tr

http://www.cucurbitaceae2012.org/invitation.html

Upcoming Meetings & News of Interest

Organization/Meeting	Dates	Location	Contact
32nd Annual Meeting of the Watermelon Research & Development Group	February 2012 8:00 am -5:00 pm	In conjunction with the 72nd Annual Meeting of the Southern Region - American Society for Horticultural Science, Birmingham, AL, USA	Jonathan Schultheis jonathan_schultheis@ncsu.edu
	October, 2012	In conjunction with Cucurbitaceae 2012, Antalya, Turkey.	
ISHS V International Symposium on Cucurbits	May 14-16, 2013	Giza, Egypt	Ahmed Glala, Hoticultural aaa_glala@yahoo.com
Cucurbit Crop Germplasm Committee Meeting	TBA	In conjunction with Cucurbitaceae 2014, USA.	Jim McCreight jmccreight@pw.ars.usda.gov
Cucurbit Genetics Cooperative Business Meeting	September 2011	In conjunction with American Society for Horticultural Science, Waikoloa, Hawaii	Todd Wehner todd_wehner@ncsu.edu
	July-August 2012	In conjunction with American Society for Horticultural Science, InterContinental Miami Hotel, Miami, Florida	
	October 2012	In conjunction with Cucurbitaceae 2012, Antalya, Turkey.	
Pickle Packers International	October 26-28, 2011	Venetian Hotel, Las Vegas, Nevada.	202-331-2456 http://www.ilovepickles.org
	April 10-12, 2012	Hyatt Regency, Milwaukee, WI	
Cucurbitaceae 2014	TBA	TBA	Rebecca Grumet grumet@msu.edu Mike Havey michael.havey@ars.usda.gov
X EUCARPIA International Meeting on Cucurbitaceae Eucarpia 2012	October 2012	Rixos Downtown Antalya Hotel, Antalya, Turkey	Nebahat Sari nesari@cu.edu.tr
Melon Breeders Group	TBA	In conjunction with Cucurbitaceae 2014, USA.	Jim McCreight Jim.McCreight@ars.usda.gov
National Watermelon Association	February 22-26, 2012	Ritz CarltonResort & Spa, Amelia Island, Florida	Telephone: 863-619-7575 Fax: 863-619-7577 nwa@tampabay.rr.com http://www.nationalwatermelonassociation.com
	February 20-24, 2013	Westin la Cantera Resort & Spa, San Antonio, Texas	
Squash Research Group	October 2012	In conjunction with Cucurbitaceae 2012, Antalya, Turkey.	TBA
Pickling Cucumber Improvement Committee	October 28, 2011	In conjunction with Pickle Packers International (PPI) Annual Meeting & Product Showcase, Venetian Hotel, Las Vegas, Nevada.	Yiqun Weng weng4@wisc.edu

Cucurbit Genetics Cooperative

Style Guide

The following *guidelines* are for use in the preparation of reports. It is recognized that CGC members may not be able to meet one or more of the guidelines.

Authors are encouraged to contribute reports even though some of the guidelines cannot be met.

Our objective is to facilitate the interchange of information, but we ask authors to help reduce unnecessary editing.

Refer to the latest Cucurbit Genetics Cooperative Report regarding questions of style not mentioned.

I. **Reports will be assigned to one of the following:**

    A. Research Notes - short reports dealing with current genetics, breeding and closely related matters that are of possible interest to members.

    B. Germplasm Exchange - a listing of seed stocks that are available or desired. Brief descriptions and gene symbols, if applicable, are useful.

II. **General Guidelines**

    A. Reports should normally not exceed two (2) single-spaced, typewritten or word-processed pages.

    B. Authors are requested to submit electronic copy of their reports by email. The report should be submitted as a word processing file. A follow up email should be sent to see if it was properly received.

    C. Tables and Figures (e.g., *.TIFF, *.PCX, *.GIF, *.JPG, *.WPG) should be included as separate files on the disk even if they are also embedded in the body of the text.

    D. If you are unable to submit your report by email or disk, send a typed copy. CGC will look after re-entering your submission.

III. **Title**

    A. The title should be a precise and concise description of the work.

    B. Avoid the use of meaningless words such as "influence of," "effects of," "results of," "studies on," "evaluation of," "factors involved in," and "tests on."

    C. Begin at left-hand margin. (See Examples I, II and III)

    D. Capitalize first letter of all words except for articles such as "a" and "the," prepositions such as "of," "in," "on," "during," and "between," and conjunctions such as "and" and "with" that are not the first word.

    E. DOUBLE SPACE between Title and By-line.

IV. **By-line**

    A. Author(s) name(s) (first name or initial followed by middle initial and last name). (See Example I)

        1. Names of two or more authors at the same institution are on the same line. (See Example II)

        2. Names of authors in separate institutions are on different lines. (See Example III)

    B. Concise mailing address is on the line below the author(s) name(s). (See Examples I, II and III)

    C. TRIPLE SPACE between By-line and Body of Report. (See Example I)

V. Body of Report (See Example I)

A. Follow conventional format and include a brief Introduction, essential Materials & Methods, and concise Results and Discussion.

B. DO NOT indent the first word of a paragraph.

C. Use numbers enclosed in parentheses for literature citations.

D. DOUBLE SPACE between paragraphs and between body of report and Literature Cited.

VI. Taxonomy and Genetic Nomenclature (See Example I)

A. <u>Taxonomy</u> (See Example I)

    1. Give the full scientific names of plants, disease organisms, and insects, along with their authority (and if important, the cultivar name).

    2. *Italicize* scientific names.

    3. Use common names whenever possible.

    4. Cultivar names can be preceded by the abbreviation for the word cultivar (e.g., cv. Calypso), or can be set off with single quotes (e.g., 'Calypso').

B. <u>Genetic Nomenclature</u> (See Example I)

    1. Names and symbols of genes are subject to the gene nomenclature rules for the Cucurbitaceae. (Robinson et al. 1976. Genes of the Cucurbitaceae. HortScience 11:554-568; CGC Gene List Committee. 1982. Update of cucurbit gene list and nomenclature rules. Cucurbit Genetics Cooperative Report 5:62-66.) These rules were reprinted in the latest CGC Report.

    2. Refer to the rules of nomenclature before assigning a name and symbol to a newly described gene in a published report regardless of where it is published.

    3. If necessary, consult the CGC Gene List Committee regarding questions of gene names and symbols. Members of the Gene List Committee are listed in the latest CGC Report.

    4. *italicize* gene names and symbols.

VII. Literature Cited (See Example I)

A. List citations in alphabetical order, but numbered consecutively with Arabic numerals followed by a period.

B. Authors are listed after the number; senior author (last name first, by initials), then additional authors (initials first).

C. DO NOT substitute the underline for the author's name when an author is cited more than once, repeat the author's name for each citation.

D. DO NOT indent the second and any subsequent lines of citations, but begin directly below the first letter of the author's last name.

E. DO NOT underline journal titles.

VIII. Tables (See Example IV)

A. Tables should document or clarify, but not duplicate, data already given in the text or figures.

B. Large tables can be reduced in size through photoreduction (or reduced font size) in order to fit within the prescribed margins. Photoreductions should be done by the author(s) if possible.

C. Table Anatomy

1. Headnote - contains "Table," then number in Arabic, and a self-explanatory title.

2. Headrule - underscores the headnote; one line.

3. Stubhead - is the head of the first column. Capitalize only the first letter of the first word and any proper nouns.

4. Boxhead - contains the column heads of the rest of the table, and is centered between the stubhead and the right margin. Capitalize only the first letter of the first word and any proper nouns.

5. Boxhead rule - one line under the boxhead to separate it from the main body of the table.

6. Field - is all the information between the boxhead rule and the footrule - - the main body of the table.

7. Footrule - a single underscore to separate the field from the footnotes (if any).

8. Footnotes - are designated with superscript, lowercase letters in reverse alphabetical order (z, y, x, w, etc.), thus avoiding confusion with alphabetical letters used for statistical significance (a, b, A, B).

IX. Figures

A. Data presented in tables should not be duplicated in Figures.

B. Figures include graphs and line drawings in black on white paper or on white paper imprinted with light blue lines which will not appear when photographically reproduced, and black and white photographs.

C. Large figures can be reduced in size through photoreduction in order to fit within the prescribed margins. Photoreductions should be done by the author(s) if possible.

D. Captions should be clear, concise and complete.

E. If mailing reports, protect figures with stiff cardboard backing and mark envelope "Do Not Bend."

Examples

Example I

Sources of Resistance to Viruses in Two Accessions of *Cucumis sativus*

R. Provvidenti

Department of Plant Pathology, New York Agricultural Experiment Station, Cornell University, Geneva, NY 14456

Recently we have determined that two accessions of *Cucumis sativus* L. cv. Surinam and cv. TMG-1 are valuable sources of resistance to the most common viruses affecting this species in the U. S.

'Surinam', a cultivar from the South American country of the same name, possesses a single gene (*wmv-1-1*), which confers resistance to watermelon mosaic virus 1 (WMV-1) (2). Following inoculation . . .

(body of report)

...breeders with sources of resistance to four viruses.

Literature Cited

1. Provvidenti, R., D. Gonsalves, and H.S. Humaydan. 1984. Occurrence of zucchini yellow mosaic virus in cucurbits from Connecticut, New York, Florida, and California. Plant Disease 68:443-446.

2. Wang, Y.J., R. Provvidenti, and R.W. Robinson. 1984. Inheritance of resistance to watermelon mosaic virus 1 in cucumber. HortScience 19:587-588.

Example II

Obtention of Embryos and Plants from In Vitro Culture of Unfertilized Ovules of *Cucurbita pepo*

D. Chambonnet and R. Dumas de Vaulx

Institut National de la Recherche Agronomique, 84140 Montfavet, France

Example III

Lack of Resistance to Zucchini Yellow Mosaic Virus in Accessions of *Cucurbita maxima*

R. Provvidenti

Department of Plant Pathology, New York Agricultural Experiment Station, Cornell University, Geneva, NY 14456

R. Alconero

U. S. Department of Agriculture, Agricultural Research Service, Regional Plant Introduction Station, Geneva, NY 14456

Example IV

Table 1. Petiole length (cm) of the first four true leaves of mutant and normal cucumber plants segregating for the short petiole (*sp*) gene.

| Genotype | Leaf node | | | |
	1	2	3	4
sp sp	1.9	1.8	6.7	3.2
Sp sp	15.0	14.2	16.2	16.1
Sp Sp	15.2	15.9	17.6	17.8

In Memoriam

Warren S. Barham, Watermelon Breeder

Dr. Warren S. Barham worked in administration, research and teaching at Cornell University, North Carolina State University, and Texas A&M University. He worked many years in the seed industry at several seed companies before founding his own company, Barham Seeds, Inc. in 1986.

Dr. Barham was the first watermelon breeder at North Carolina State University. A major contribution to watermelon breeding at NC State was finding resistance to anthracnose (*Colletotrichum lagenarium*). He also studied physiological disorders in watermelon.

Warren Barham studied seed size in watermelon and developed the small (*short*) seed type, which was released in the cultivar 'Sweet Princess'. He studied the inheritance of golden leaf and fruit (*go*) as a possible indicator of ripeness.

While in the seed industry, Dr. Barham and his staff developed over twenty seedless and seeded watermelon hybrids. A&C 5244 is one of the most popular seedless watermelons grown in North America. Along with watermelon breeding, he also bred onion, tomato, and cucumber.

Warren S. Barham, PhD, passed away on April 16, 2010. He was born February 15, 1919, in Prescott, Arkansas. He married Margaret Alice Kyle (deceased 1997) in 1940. Warren and Margaret had 4 children (Barbara, Juanita, Margaret Ann, and Robert; 11 grandchildren; and 11 great-grandchildren).

Warren was a generous loving father, husband, grandpa, and friend with a wonderful sense of humor and compassion for all. He knew no stranger and had an amazing ability to keep in touch with friends and colleagues from his native Arkansas, to all corners of the world. He was a passionate horticulturist, educator and renowned plant breeder who never retired from his creative instincts to develop new fruits and vegetables. He had a strong, lasting influence on all who knew him.

Dr. Barham received his BS degree at the University of Arkansas, Fayetteville, in 1941 and his PhD in Plant Breeding, Vegetable Crops, and Plant Physiology at Cornell University, Ithaca, N.Y. in 1949. Dr. Barham served in the Army Air Corps during World War II from 1942 to 1945. Dr. Barham was Assistant Professor of Horticulture at North Carolina State University, Raleigh, where he taught Plant Breeding and Graduate Research until 1958. He then became Director of Raw Products Research for Basic Vegetables in Vacaville,

California, until 1976, and was then Professor and Head of the Horticultural Sciences Department at Texas A&M University, College Station, until 1982.

Dr. Barham moved to California and formed Barham Seeds, Inc. in 1985 After "retiring" for two weeks, at 65, Warren Barham then went on to become a pioneer in the development of seedless watermelons.

Dr. Barham continued active research, consulting, and generating innovative ideas about vegetable breeding until the final day of his life. Dr. Barham's research included identifying and developing solutions to long-standing issues for plant breeding in various crops including onions, tomatoes, watermelons, melons, and cucumbers.

Dr. Barham was committed to contributing to his community as well as to his profession. He served as a school board member in Vacaville for 10 years, as a Rotarian for over 50 years, served as a member of parks and recreation commission in Gilroy, and volunteered at the Garlic Festival. He backed up his time commitment to community service with generous donations to fighting cancer, supporting the arts, helping youth sports, scholarships, and public horticultural research – to name a few.

Significant was Dr. Barham's commitment to the American Society for Horticultural Science (ASHS). He was always supportive and active in this preeminent scientific society. He was vice-president of Industry and then president of ASHS.

Warren Barham is an excellent example of that "great generation" that first saved the world for freedom then went on to enhance life for everyone through hard work and positive achievement.

Henry Munger, Plant Breeder Extraordinaire

R. W. Robinson
Horticulture Dept., Cornell University
Geneva, NY 14456

Henry M. Munger died on August 25, 2010 at the age of 94. He was a renowned breeder of many vegetable crops. He introduced more than 70 varieties and breeding lines of nine different vegetable crops.

Dr. Munger grew up on a vegetable farm in western New York and entered Cornell University at the young age of 16. After receiving his BS from Cornell in 1936 he obtained his MS from Ohio State University in 1937, and then returned to Cornell where he was awarded his PhD degree in 1941. After graduating he was a faculty member at the University of Wisconsin for two years, then returned again to Cornell University where he had joint appointments in the Dept. of Plant Breeding and the Dept. of Vegetable Crops.

Improved quality and disease resistance were of prime importance in his many breeding programs. He diligently tasted the fruit of hundreds of tomato breeding lines in a quest to breed for better flavor. The 'Gardener' and 'VF Gardener' varieties that he bred have exceptional flavor, and he bred three other tomato varieties as well. He bred onions for mild flavor and long storage life and, in cooperation with other breeders, developed the 'Empire', 'Premier', and 'Sweet Sandwich' varieties. He bred cabbage for yellows resistance and introduced the 'Empire Danish' variety. He bred the blight resistant celery varieties 'Emerson Pascal' and 'Beacon'. He also bred spinach, beans, peas, and *Amaranthus*. The male sterile mutant of the Queen Anne's Lace weed which he discovered while on vacation on Cape Cod is the basis for all hybrid carrot varieties and the important baby carrot industry.

Dr. Munger was an outstanding breeder of many vegetable crops but cucurbits were his favorite. Melon was the first crop he ever bred and at the end of his long and distinguished career he was still breeding melon, and cucumber and squash as well.

His doctoral research, under famed plant breeder R. A. Emerson, was on Fusarium resistance for melon. This led to his development of 'Iroquois', the first Fusarium resistant variety of melon, and later to 'Delicious 51'. He also bred melons for resistance to powdery mildew and cucumber mosaic virus. He developed an improved method for producing hybrid melon seed, and he bred monoecious melons with round fruit shape so that seedsmen could produce hybrid seed without having to emasculate bisexual flowers.

His accomplishments in cucumber breeding are legendary. His 'Marketmore' variety, the improvements he added to it by backcrossing, and varieties with 'Marketmore' in their pedigree that were bred by others represent most of the cucumber varieties grown in the US and many other countries today. He bred cucumber varieties with a higher level of CMV resistance than any before, with combined resistance to more diseases than any other variety, and with better and more uniform fruit color. Cucumber varieties he bred include 'Yorkstate Pickling' (1950), 'Niagara' (1951), 'Tablegreen' (1959), 'Marketmore' (1968). 'Meridian' (1971). 'Marketmore 76' (1976), 'Poinsett 76' (1976), 'Spacemaster' (1980), and 'Comet' (1983).

Henry was the foremost proponent of the use of backcrossing in vegetable breeding. Other vegetable breeders have used the backcross method to incorporate single genes into varieties, but none so extensively as Dr. Munger. No one else to my knowledge has used the backcross method with traits of complex inheritance, but Dr. Munger accomplished this with CMV resistance for cucumber. It is difficult to identify cucumber plants with all possible CMV resistance genes when evaluating the selfs of each backcross generation, since plants lacking one of the genes may have nearly as high a level of CMV resistance as those homozygous for all of the resistance genes. Dr. Munger, however, determined that backcross plants not selfed, which are heterozygous for the resistance genes, have an intermediate level of resistance and plants with all of the resistance genes can be more easily distinguished.

Dr. Munger backcrossed resistance to the phenomenal number of nine diseases (CMV, ZYMV, WMV, PRSV, scab, powdery and downy mildew, bacterial wilt, and target leaf spot) into 'Marketmore' and other cucumber varieties he bred and also into 'Wisconsin SMR 18', and 'Poinsett'. He also backcrossed genes for nonbitterness and cucumber beetle resistance, plant habit, and uniform fruit color. The germplasm that he developed by backcrossing has been very valuable in cucumber breeding. They provide breeders with parents for multiple disease resistance and other useful traits. 'Marketmore 76' and 'Poinsett 76', which he developed by backcrossing, have become important varieties. His gynoecious versions of 'Marketmore', 'Tablegreen', and 'Poinsett' have value as the female parent of hybrid varieties. His near isogenic lines make possible basic physiological research to investigate the effect of a gene of horticultural importance without its being affected by other genes.

Henry Munger bred 'Butternut 77', a winter squash variety of *Cucurbita moschata* with reduced vine size. He also bred *C. moschata* for disease resistance. Henry Munger was the first to breed disease resistant squash, and almost all of the disease resistant varieties of squash and pumpkin now being grown are derived from germplasm he bred.

It is time consuming to inoculate cucurbits by the method previously used to breed for virus resistance, by manually rubbing inoculum onto leaves of individual plants. This problem was overcome, however, when Dr. Munger developed an ingenious method of using a leaf blower to project inoculum with considerable force into many plants at the same time. This method has expedited the development of disease resistant squash and pumpkin varieties.

Previously, all attempts to find a useful source of disease resistance in any of the cultivated species of *Cucurbita* were unsuccessful. Henry therefore asked

Tom Whitaker if any of the compatible wild species could be used as a source of powdery mildew resistance, and he was advised to use *C. okeechobeensis* subsp. *martinezii* as a parent. Dr. Munger used this species to breed disease resistance into both winter and summer squash, and made this germplasm freely available to other breeders.

It was difficult for him to use *martinezii* as a parent since it does not flower in the field at Ithaca, NY until shortly before frost, but Henry succeeded in crossing it with 'Butternut'. It is even more difficult to cross *martinezii* with summer squash due to severe sterility barriers with *C. pepo* but Dr. Munger and his graduate student, Max Contin, succeeded in crossing summer squash with the hybrid of *moschata* x *martinezii*. They thereby used *moschata* as a bridge in order to introgress disease resistance from *martinezii* to *pepo*.

He used *martinezii* to breed squash for powdery mildew resistance, and then his keen power of observation led him to discover that some of his breeding lines were also resistant to naturally occurring CMV. His finding that *martinezii* is resistant to CMV encouraged his colleague at Geneva NY, Rosario Provvidenti, to test other wild *Cucurbita* species and to find sources of resistance to many viruses. Based on this information, Henry then used *C. ecuadorensis* as a parent for resistance to ZYMV, WMV, and PRSV. When Dr. Provvidenti later discovered that Nigerian Local (*C. moschata*) is resistant to the same viruses, Henry used this more tractable parent for breeding summer squash with multiple virus resistance.

Dr. Munger received many well deserved honors and awards. He was profiled in a well written biography by Martha Mutschler in an issue of Plant Breeding Reviews dedicated to him. He served as President of the American Society for Horticultural Science. He was the first living recipient of the Horticultural Hall of Fame of the ASHS, which previously included only such eminent authorities as Liberty Hyde Bailey, Luther Burbank, and Gregor Mendel. ASHS also designated him as a Fellow of that society and honored him with the Norman F. Childers Award for distinguished graduate teaching. He was awarded the prestigious World Seed Prize. The University of Nebraska honored him with an honorary doctorate degree. Dr. Munger was given the Man of the Year award by the Vegetable Growers Association of America. His contributions to vegetable breeding were also honored by the New York Vegetable Growers Association and by the National Council of Plant Breeders. He was given the All America Selections award for outstanding achievements in horticulture. In 1989 the H. M. Munger Symposium on Breeding Vegetables for Virus Resistance was held at Cornell University, and many of his former students returned to Cornell to attend it and honor him.

He served as editor of the Proceedings of the American Society for Horticultural Science. He was the founder and the editor of the Vegetable Improvement Newsletter, the precursor of the Cucurbit Genetics Cooperative Reports. He served as Head of the Department of Vegetable Crops at Cornell University for 16 years.

I was privileged to have Henry as a teacher, a mentor, a colleague, and as a friend.

The *Cucumis* of Antiquity: A Case of Mistaken Identity

Harry S. Paris
Department of Vegetable Crops & Plant Genetics, Agricultural Research Organization, Newe Ya'ar Research Center, P. O. Box 1021, Ramat Yishay 30-095, Israel

Jules Janick
Department of Horticulture & Landscape Architecture, Purdue University, 625 Agriculture Mall Drive, West Lafayette, IN 47907-2010, U.S.A.

It has become nearly axiomatic that "cucumbers" were familiar to ancient Mediterranean civilizations. A number of authors specializing in cucurbits, almost a Who's Who of Cucurbitology, have written that the cucumber, *Cucumis sativus* L., spread westward from its homeland in the foothills of the Himalayas some 3000 years ago. It is also understood axiomatically that since the English word "cucumber" looks so much like the Latin *cucumis*, that they must be one and the same. Upon closer examination, it can be seen that this "fact" is baseless.

The Latin *cucumis* is indeed the source of the modern English word "cucumber" and translators into English of the works of Columella and Pliny, 1st-century Roman authors who wrote in Latin, used "cucumbers" for *cucumis*. However, Columella and Pliny described the *cucumis* as snake-like and hairy (4). The fruits of cucumber, *Cucumis sativus* L., are glabrous but the young fruits of melon, *C. melo* L., are hairy. A partially preserved fresco at the ruins of Ercolano (Herculanum), a city destroyed along with Pompeii during the eruption of the Vesuvius volcano in 79 CE, shows several striped snake melons, much like 'Armenian Striped', inside and next to a large glass jar, indicating that snake melons were not only eaten fresh, but also were pickled.

Hebrew writings by 2nd- and 3rd-century Jewish authors mention the *qishu'im* a number of times and this word has also been translated into English as cucumbers. In these writings, too, the *qishu'im* are described as hairy and, moreover, as having to undergo *piqqus*, removal of the hairs, in order to be fit for eating (7). Again, the hairiness of the young fruits indicates that melons, not cucumbers, are being discussed.

Greek writings allude to *sikyos*, which has also been translated as cucumbers. These writings go back to the *Regimen* of Hippocrates, ca. 400 BCE (5). More telling, though, is the description of the *sikyos* by Theophrastus, ca. 300 BCE (3). Theophrastus was a systematic botanist as he attempted to classify plants by their distinguishing features. He described the *sikyos* as an herbaceous plant that has a long period of bloom. Its fruit is made of flesh and fiber and the seeds within are arranged in rows. The flowers persist for a long time while the fruits are developing. Thus far, the description could fit very well both *Cucumis sativus* and *C. melo*. But he also stated: *Some flowers are sterile, as in* sikyon, *those which grow at the ends of the shoot, and that is why men pluck them off, for they hinder the growth of the* sikyoi. By sterile flowers, of course, he would be alluding to the staminate flowers and, according to the description, these are borne on the end of the shoot. Plants of *C. sativus* become increasingly pistillate as they develop (9). In sharp contrast, plants of *C. melo* bear pistillate or hermaphroditic flowers only on the first one or two nodes of branches, all apical nodes are staminate (8). Hence, the description by Theophrastus was of *C. melo*, not *C. sativus*.

Ancient Egyptian wall paintings of elongate fruits have been interpreted as cucumbers, but the striping of some fruits and the furrowing of some others is more consistent with melons (4).

Obviously, the evidence from four Mediterranean civilizations, Egyptian, Greek, Jewish, and Roman, agrees that vegetable melons, mostly chate melons early on (Egypt) and snake melons later on (Greece, Israel, and Rome), were valued and familiar to them.

There is also evidence for the use of snake melons across the Mediterranean and into the Middle East during the medieval period. They are still widely grown today in the warmer regions of the Old World, from northern Africa to India (1,2,6). They do not thrive in cooler climates, however, so when *Cucumis sativus* was finally spread westward from India, it was probably welcomed and the derivatives of the word *cucumis* were expropriated to this species.

The translation of the classical Latin *cucumis* as cucumber may be acceptable to the general public but is problematic when extended to use by students of botany, horticulture, and crop history. It may be quixotic to try to erase the oft-repeated "fact" that the spread westward of cucumbers from the Indian subcontinent to Mediterranean civilizations occurred at least 3000 years ago, but as researchers specializing in cucurbits we cannot escape our responsiblity to communicate accurately and attempt to correct a past wrong. There is no evidence for the arrival of cucumbers in Mediterranean lands prior to 1500 years ago. We have found evidence indicating that *C. sativus* arrived in this area early in the medieval period, about the time of the Islamic conquests.

Literature Cited

1. Chakravarty, H.L. 1966. Monograph of the Cucurbitaceae of Iraq. Technical Bulletin 133, Ministry of Agriculture, Baghdad.

2. Hassib, M. 1938. Cucurbitaceae in Egypt. Noury & Fils, Cairo.

3. Hort, A. 1976. Theophrastus, enquiry into plants, 2 vol. London: William Heinemann.

4. Janick, J., H.S. Paris, and D.C. Parrish. 2007. The cucurbits of Mediterranean antiquity: Identification of taxa from ancient images and descriptions. Ann. Bot. 100: 1441–1457.

5. Jones, W.H.S. 1967. Hippocrates, vol. 4. London: William Heinemann.

6. Pandey, S., N.P.S. Dhillon, A.K. Sureja, D. Singh, and A.A. Malik. 2010. Hybridization for increased yield and nutritional content of snake melon (*Cucumis melo* L. var. *flexuosus*). Plant Genet. Resourc. 8: 127–131.

7. Paris, H.S. and J. Janick. 2008. Reflections on linguistics as an aid to taxonomical identification of ancient Mediterranean cucurbits: The piqqus of the faqqous. In: M. Pitrat, ed., Cucurbitaceae 2008, pp. 43–51. I.N.R.A., Avignon, France.

8. Rosa, J.T. 1924. Pollination and fruiting habit of the cantaloupe. Proc. Amer. Soc. Hort. Sci. 21: 51–57.

9. Shifriss, O. 1961. Sex control in cucumbers. J. Hered. 52: 5–12.

Origin and Characterization of the 'Lemon' Cucumber

R. W. Robinson

Horticulture Dept., Cornell University, Geneva, NY 14456

A touching tale was told by a huckster in the year of 1909 (3). In order to commemorate the wedding of his beloved daughter, he claimed that he plucked an orange blossom from her bridal bouquet and used it to pollinate a cucumber plant. The result of this bizarre union was, he said, a cucumber plant bearing fruit like that of an orange. He described its flesh as having the most delicious blending of the finest Florida orange with the crispness of a delectable cucumber. He offered to share his marvelous creation with others for only a dollar for each seed, which of course would be much more expensive with the inflated dollars of today. This preposterous claim of a cucumber x orange intergeneric hybrid was indignantly denounced by Fullerton (3) in 1910. He disclosed that it was nothing more than the 'Lemon' cucumber cultivar, which was listed in seed catalogs at that time.

Fruit of this cucumber cultivar does bear a slight resemblance to a citrus fruit. It is nearly round, not elongated like most cucumbers, and about the size of a small orange. The fruit often has a protrusion at the blossom end, somewhat like that of a navel orange. The fruit color changes during development from a light green to a pale yellow, like a lemon, and then to a deeper yellow with faint orange mottling.

The first record I found of 'Lemon' cucumber in the extensive seed catalog collection of the Bailey Hortorium at Cornell University was in the 1894 catalog of Samuel Wilson, which includes this description: "They are considered a great dainty by those who are fond of that popular vegetable. They have all the desired qualities of a good cucumber, slightly flavored with lemon, which gives them a decided advantage over the common kind." The 1901 seed catalog of James Vick & Sons offered a packet of 'Lemon' seed for a dime and described it thus: "The flesh is exceedingly tender and crisp, with a sweet flavor surpassing all other cucumbers. They have none of the bitter or acid taste so generally found in cucumbers". Other early listings of 'Lemon' in seed catalogs include L. L. May & Co. (1902), John Gregory & Sons (1902), J. E. Thornburn (1907), Iowa Seed Co. (1909), Aggeler & Musser (1909), and A. W. Livingston & Sons (1910). 'Lemon' is still listed in some seed catalogs today. It may be the oldest cucumber cultivar now being grown.

Some home gardeners grow it for the novelty of its round, yellow, nonwarty fruit. 'Lemon' cucumbers are occasionally seen in farmers markets but the cultivar has no real commercial importance. It is considered by some to have a better flavor than conventional cucumbers, being nonbitter and very sweet. Its reported freedom from bitterness and indigestion problems may be due to a low cucurbitacin content. 'Lemon' fruit have a high soluble solids content. It had 5.4% Brix, the highest of any of the 245 cultivars, plant introductions, and breeding lines that were tested with a hand refractometer (10). Soluble solids content is a better measure of sweetness in melons and watermelons than in cucumber. Cucumber fruit have little sucrose but rather have reducing sugars as a major component of soluble solids (6).

'Lemon' is late in maturity, low yielding, and has no disease resistance. It has been of value, however, for genetic research. Its andromonoecious sex expression, with male and female elements in the same blossom, results in more natural self pollination than that of monoecious cucumbers which have the sexes in separate flowers. This makes it feasible to use open pollination to obtain recessive mutants of rare occurrence more efficiently than with hand pollination. Using 'Lemon' with its relatively high rate of natural self pollination made it possible for me to use open pollination to obtain *gl, cor-2, dw, lh* and many other cucumber mutants after treatment with thermal neutron radiation.

'Lemon' has also been useful in cucumber breeding. Kubicki (5) determined that hermaphroditic plants with the genotype *m/m F/F* occur in the F-$_2$ generation when 'Lemon' is crossed with a gynoecious parent. This has made possible the development of homozygous gynoecious hybrid cultivars (*m/+ F/F*) that are more stable in gynoecious sex expression than hybrids heterozygous for *F* (9).

Tkachencko (12) hoped to use 'Lemon' as a parent to breed a greenhouse cucumber that would not require hand or insect pollination to set fruit because he considered that its bisexual flowers would naturally self pollinate, but he was not successful. If the perfect flower is not insect or hand pollinated it generally does not set fruit. The pollen is shed on the outside of the anthers, away from the central pistil, and does not usually reach the stigma unless brought there manually or by bees.

Tkachencko (12) considered 'Lemon' to be so different from other cucumbers that he designated it as a distinct species, *Cucumis sphaerocarpus*, named after its round fruit. But it is *C. sativus*, of course, and only a few genes determine the major characteristics that distinguish 'Lemon' from other cucumber cultivars (13). Its andromonoecious sex expression is due to gene *m* and this gene is associated with nearly round fruit shape. The fruit shape of 'Lemon' is also influenced by the *l* gene for five fruit locules instead of the normal three. Its yellow-green immature fruit color is governed by *yg* . Gene *n* is responsible for the upright rather than pendant position of immature 'Lemon' fruit. Its protruding ovary is due to *pr*. The opposite leaf arrangement is determined by *opp* and the fasciation of some 'Lemon' plants is due to *fa*. All of these genes are recessive and when 'Lemon' is crossed with a white spined monoecious cucumber the only outward evidence of 'Lemon' in the hybrid is its black fruit spines.

Many of the characteristics that make 'Lemon' unique are genetically connected by linkage or pleiotropy. Andromonoecious sex expression, negative geotropic peduncle response, round fruit shape, protruding ovary, opposite leaves, and fasciation are completely linked or are pleiotropic. Genes for more than three fruit locules and for yellow fruit color are also on the same chromosome but are less closely linked with the andromonoecious sex expression gene.

Some of the unique characteristics of 'Lemon' cucumber may have existed since antiquity. Chakravarty (2) stated that a Sumerian cucumber grown in Iraq thousands of years ago had short ovoid or round fruit. In 1859 Naudin (7) wrote of a land race of cucumber from India having fruit with five locules instead of the customary three. A cucumber plant with opposite leaf arrangement was portrayed in an ancient herbal (8) and cucumbers with fruit shape like 'Lemon' were depicted in herbals of the 16th century (4).

Almost 2000 years ago Pliny the Elder wrote about a cucumber shaped like a quince and golden in color, a description resembling the fruit of Lemon' (1). But it likely was not 'Lemon' or any other cucumber since Pliny stated that its fruit separated from the vine when mature. Fruit dehiscence is characteristic of *C. melo*, not *C. sativus*. 'Lemon' cucumber is also distinct from the Vine Peach, *Cucumis melo* var. *chito*, which has been known as 'Lemon Cucumber' (11).

The cultivar 'Lemon' apparently originated in the late 19th century. An unresolved question is its parentage and where its unique combination of associated characteristics is derived from.

Literature cited

1. Bostock, J. and H. T. Riley. 1886. The Natural History of Pliny. Vol. 4. Henry Bohn, London. 523 pp.

2. Chakravarty, H. L. 1966. Monograph on the Cucurbitaceae of Iraq. Iraq Ministry of Agriculture Tech. Bull. 133.

3. Fullerton, H. B. 1910. Fakes and facts. Long Island Agronomist 3 (13): 10.

4. Janick, Jules and Harry S. Paris. 2006. The cucurbit images (1515-1518) of the Villa Farnesina, Rome. Ann. Bot. 97: 165-176.

5. Kubicki, B. 1965. New possibilities of applying different sex types in cucumber breeding. Genet. Polonica 6: 241-250.

6. McCombs, C .L., H. N. Sox, and R. L. Lower. 1976. Sugar and dry matter content of cucumber fruits. HortSci. 11: 245-247.

7. Naudin, C. 1859. Especes et des variates du genre *Cucumis*. Ann. Ser. Nat. Ser. 4. 11:5-87.

8. Paris, H. 2006. Oral presentation at Cucurbitaceae 2006 Meeting, Asheville NC.

9. Robinson, R. W. 1999. Rationale and methods for producing hybrid cucurbit seed. Jour. New Seeds 1: 1-47.

10. Robinson, R. W. 1987. Genetic variation in soluble solids of cucumber fruit. Cucurbit Genet. Coop. Rpt. 10: 9.

11. Tapley, W. T., W. D. Enzie, and G. P. van Eseltine. 1937. The Vegetables of New York. Vol. 1, Part IV. J. B. Lyon Co., Albany NY.

12. Tkachenko, N. N. 1925. Preliminary results of a genetic investigation of the cucumber – *Cucumis sativus* L. Bull. Appl. Pl. Breed. Ser. 2, 3: 311-356.

13. Youngner, V. B. 1952. A study of the inheritance of several characters in the cucumber. PhD thesis. Univ. Minnesota.

Yield of Spring-Planted Cucumber Using Row Covers, Polyethylene Mulch, and Chilling-Resistant Cultivars

Todd C. Wehner, Gabriele Gusmini and Katharine B. Perry
Department of Horticultural Science, North Carolina State University, Raleigh, NC 27695-7609

North Carolina is a leading producer of field-grown cucumbers (*Cucumis sativus* L.) in the United States. In the 2001 to 2003, North Carolina ranked second in the production of processing (pickling) cucumbers, after Michigan, with approximately 74,700 Mg harvested per year. In the same period, North Carolina ranked fifth in the production of fresh-market (slicing) cucumbers (50,000 Mg per year) after Florida, Georgia, California, and Michigan (5).

In North Carolina, growers produce a spring and a summer crop. The primary production area is the coastal plain, where the spring crop is planted mid-April, approximately one month earlier than in the mountains, the secondary production area (3). Strategies to extend the production season of cucumbers in environments where chilling injury of the seedlings may occur, include the development of chilling resistant cultivars that can germinate at low soil temperatures and are resistant to chilling (low temperatures above freezing). In a study of environmental effects on response to chilling treatments in cucumber, chilling resistance was determined by growth temperature before chilling, chilling temperature and duration, light intensity during chilling, and genotype (4). Based on these results, the USDA-ARS cucumber germplasm collection was screened to rank PI accessions, cultivars, and breeding lines for resistance to chilling injuries (Smeets and Wehner, data not shown).

The use of polyethylene mulches in horticulture has been widely adopted for the control of weeds and the reduction of herbicide use (1). Polyethylene mulches are applied to cover raised beds, after incorporation of herbicides, while the soil between beds is kept weed-free through cultivation and herbicide applications. In addition, fumigation under the mulch strips may help to control weeds, although fumigations are done mostly to control soil-borne pathogens and nematodes.

Polyethylene mulches may also affect yield in horticultural crops, by increasing the soil temperature and the amount of light reflected from the soil onto the canopy. For example, the mulch surface color had a significant effect on total yield and earliness of fresh-market tomatoes, by influencing the plant microclimate and stimulating higher and earlier fruit production in this crop. The comparison of red, black, silver, and white mulch colors resulted in higher yields from plots with red or black mulch (2).

Row covers are commonly used for the production of horticultural crops when the average temperatures during the growing season are lower than the optimum for plant growth. There are two major types of row covers: polyethylene slitted film mounted on wire hoops and floating polyester. The major advantage of the second type is the easier installation system with a modified polyethylene mulch applicator. Even though these two types of row covers offer the best level of control for day-time temperatures, the night-time protection from frost that they provide is not as useful. Furthermore, the humidity level underneath the row covers determines the usefulness of these materials in different environments (6). The combination of mulch and row covers allows the improvement of soil and air temperatures, as well as weed control under the row covers. There is no need to remove the covers after their placement until the end of the protective treatment.

In our study of early production of pickling and slicing cucumbers in North Carolina, we verified the effect of black polyethylene mulch, clear slitted polyethylene or floating polyester row covers, and genetic resistance to chilling. Our objectives were: 1) to determine the best combination of mulch and row cover types for early production of chilling resistant and susceptible cucumbers, and 2) to evaluate a diverse group of chilling resistant and chilling susceptible cucumber cultivars and breeding lines in early spring production in North Carolina.

Methods

We conducted our experiments at the Horticultural Crops Research Station at Clinton, North Carolina. In 1987, we used two cultivars to evaluate the best combination of mulch and row cover for early production of cucumbers in North Carolina. 'Wisconsin SMR 18' was resistant to chilling, while 'Poinsett 76' was susceptible. In 1988, we used the best combination of mulch (black polyethylene) and row cover (floating nonwoven polyester) to trial a total of 14 pickling and 14 slicing cucumber cultivars and breeding lines for early production.

'Albion', 'Calypso', 'Castlepik', 'Chipper', Gy 14A, H-19, M 21, M 28, M 29, 'Pixie', 'Raleigh', 'Wisconsin SMR 18', 'Sumter', and 'Wautoma' were pickling cucumber cultivars. 'Ashley', 'Centurion', 'Dasher II', 'Early Triumph', 'Lemon', 'Mekty Green', 'Marketmore 76', 'Poinsett 76', 'Pacer', 'Palomar', 'Sprint 440S', 'Straight 8', 'Supergreen', and 'Tablegreen 65' were slicing cucumber cultivars. The chilling resistance of the cultivars used in this study was determined by Smeets and Wehner in previous experiments (unpublished data, personal communication).

We direct sowed on raised, shaped beds on 1.5 m centers. Plots were 6.1 m long, with 0.6 m between hills, and 2.5 m alleys at each end of the plot. The experiments were conducted using horticultural practices recommended to the growers by the North Carolina Extension Service (3). Soil type at Clinton was an Orangeburg loamy sand (Fine-loamy, kaolinitic, thermic Typic Kandiudults). Field preparation included the soil incorporation of 90-39-74 kg•ha⁻¹ (N-P-K) of fertilizer, with an additional 34 kg•ha⁻¹ of nitrogen applied at vine tip-over stage. We irrigated the plots when needed for a total of 30±10 mm of water per week. We applied a tank mix of 2.2 kg•ha⁻¹ of naptalam and 4.4 kg•ha⁻¹ of bensulide for weed control.

In 1987, we sowed the plots at two early planting dates (3 and 24 March) and at the recommended date for commercial growers in North Carolina (13 April). In 1988, we sowed the plots at two early planting dates (17 March and 4 April, respectively). In 1987, we sowed 120 seeds per plot, to be thinned to 60 plants per plot. Nevertheless, none of the plots had full-stand. In 1988, we sowed 100 seeds per plot and thinned them to 80 seedlings at the two true-leaf stage.

In 1987, we used black polyethylene mulch (hereafter referred to as mulch) and compared its effect with cultivation on bare ground (hereafter referred to as none). We tested row covers made of clear slitted polyethylene on wire hoops (hereafter referred to as clear) or floating nonwoven polyester (hereafter referred to as polyester) against no row covers (hereafter referred to as open). In 1988, we trialed cucumber cultivars for early production using a combination of mulch and polyester row covers.

We harvested the plots eight times, twice per week, 1987 (19 May through 15 June) and six times, twice per week, 1988 (26 May through 16 June) for fruit yield measurements. We counted and weighed cull and marketable fruit for each plot. Yield was measured as total, marketable, and cull weight (Mg•ha⁻¹) and number (thousands•ha⁻¹) of fruit by summing plot yields over harvests.

We monitored air and soil temperatures in the ex-perimental fields with copper-constantan thermocouples attached to a micrologger and multiplexer. Air temperature sensors were placed in wooden radiation shields approximately five cm above the bed surface. The soil temperature sensors were buried ten cm deep in the soil in the center of the plot. The micrologger recorded hourly averages of the mean, maximum, and minimum of five-minute temperature readings.

We conducted statistical analyses using the MEANS, CORR, and GLM procedures of SAS-STAT Statistical Software Package (SAS Institute, Cary, North Carolina). The experiments were randomized complete block designs with four replications and a split-split-plot treatment arrangement. Factors were: planting date as whole plot, crop (pickling vs. slicing type) as sub-plot, mulch and row cover (1987) or cultivar (1988) as sub-sub-plot. In 1987, plant stand was calculated as percent of the best stand, which was obtained for both crops using black polyethylene mulch and clear polyethylene row covers at the latest planting date (13 April). In 1988, plant stand was uniform and plants were thinned to 80 per plot.

Results

In 1987, the daily air and soil temperature were similar in plots without mulch (Figure 1) and in plots with mulch (Figure 2). Plots with hoop or floating row covers had higher air temperatures than plots without row cover. At the first planting date (2 March), the average daily air and soil temperatures were 7 to 12°C. In the later two planting dates, the temperatures were consistently above 10 and 15°C, respectively.

The highest plant stand per plot was recorded at the latest (commercial) planting date in plots with mulch and clear row cover (Table 1). The mean plant stand for this treatment combination was 40, thus we considered this value as 100% stand in order to standardize proportionally the counts from the other treatments. In general, we recorded higher plant stands in plots with row covers, with the exception of plots sown on 2 March and protected with clear row covers. We did not find consistent differences in plant stand between chilling resistant and chilling susceptible germplasm. Nevertheless, the lack of full stand counts on plots sown at the commercial planting date (13 April) may indicate that factors other than chilling resistance, mulch, and row cover may have influenced plant stand in our experiment, resulting in reduced seed germination and seed vigor due to wet soil after spring rainfalls.

In 1987, total yield was increased by the use of chilling resistant germplasm, with a 46% average gain over the mean yield of chilling susceptible germplasm

(Table 1). The percent early yield at the third harvest was also higher (140%) for the chilling resistant germplasm. Overall, the use of mulch and row cover increased yield, and yields of protected plots were higher at the earlier planting dates. The higher yields at the two planting dates in March were not significantly different, but they were more than one LSD interval apart from the highest yields of plots sown at the commercial planting date of mid-April. There were no significant differences in yield for different row cover types within mulch treatment at the later planting date, but some treatments with row cover at the earlier planting dates had significantly higher yields.

In 1987, we confirmed the usefulness of mulch and row covers in increasing yield, particularly when chilling resistant and chilling susceptible cucumbers are planted earlier than typically done by commercial growers in North Carolina. Polyester row covers had a significant advantage over clear row covers only at the 2 March planting date. Total yield in plots covered with polyester was 128% higher for the chilling resistant cultivar, and 191% higher for the susceptible one. In addition, polyester row covers were easier to place on the plots and could be used for more than one production cycle. Thus, for 1988, we chose to use a combination of black polyethylene mulch and floating polyester row covers for our trial of several chilling resistant and chilling susceptible cultivars of pickling and slicing cucumbers.

In 1988, Gy 14A, 'Calypso', 'Castlepik', 'Raleigh', and M 29 for the pickling type, and 'Supergreen', 'Dasher II', 'Centurion', and 'Sprint 440S' for the slicing type, all planted at the earlier planting date, had the highest yield in the trial (Table 2). The earlier planting date greatly increased yield of the best cultivars in the trial. The top performing cultivars of the pickling crop had a 53% average gain in total yield over the same cultivars planted later. The highest-yielding of the slicing type had a 22% average gain, with the exception of 'Centurion' (gain = 3%).

The highest-yielding cultivars, planted on 17 March, had also the highest early yield at the third harvest (Table 2). The non-marketable yield (cull fruit weight) was not significantly affected by the planting date for any of the cultivars tested.

Genetic resistance to chilling seemed to favor the establishment of a better plant stand in 1987. However, it did not contribute to stand establishment or to yield improvement in 1988 since we obtained 100% plant stand in every plot. Under this conditions, cultivars that were described as chilling susceptible produced similar yields to chilling resistant cultivars within the same LSD intervals. Thus, we were not able to confirm with certainty whether chilling resistance had an advantage over susceptibility for the anticipated production of spring-planted cucumbers.

We found that the use of mulch and polyester row covers would allow early production of cucumbers (pickling and slicing types) in North Carolina. The field could be planted as early as mid-March, thus anticipating traditional cultivation of one month. Furthermore, the use of mulch and row covers in our experiment increased yield of commercial cultivars dramatically, when compared to the yield of the same cultivars planted at the commercial planting date for this crop in our state.

Further investigation is needed to determine the economics of protected culture of cucumbers for growers in North Carolina. The average market value for early production should be determined and the profit gain compared with the higher costs due to the use of polyethylene and polyester (cost of purchase, management, and disposal). Finally, a higher level of genetic chilling resistance would be useful for crops planted in early spring.

Literature Cited

1. Bonanno, A. R. 1996. Weed management in plasticulture. HortTechnology 6:186-189.

2. Decoteau, D. R., M. J. Kasperbauer, and P. G. Hunt. 1989. Mulch surface color affects yield of fresh-market tomatoes. Journal of the American Society for Horticultural Science 114:216-219.

3. Sanders, D. C., ed. 2005. Vegetable crop guidelines for the Southeastern U.S. 2005. Raleigh, North Carolina: North Carolina Vegetable Growers Association. 225 pp.

4. Smeets, L., and T. C. Wehner. 1997. Environmental effects on genetic variation of chilling resistance in cucumber. Euphytica 97:217-225.

5. USDA-ARS, Statistics of vegetables and melons, in Agricultural Statistics, W. U. S. Dept. Agr., D.C., Editor. 2004, U.S. Department of Agriculture. IV.1-IV.36.

6. Wells, O. S., and J. B. Loy. 1985. Intensive vegetable production with row covers. HortScience 20:822-826.

Table 1. Yeild of chilling resistant and chilling susceptible cucumber cultivars in early spring using row covers and plastic mulch at Clinton, North Carolina 1987.

Planting date	Soil mulch[6]	Row cover[7]	% Stand[2]		Total Mg[3]		% Cull[4]		% Early[5]	
			Res[8]	Sus[9]	Res[8]	Sus[9]	Res[8]	Sus[9]	Res[8]	Sus[9]
2 March	Mulch	Clear	13	5	14.7	5.4	15	8	49	31
		Polyester	55	30	33.6	15.7	20	9	64	26
		Open	53	3	15.8	1.1	13	20	49	10
	None	Clear	70	8	34.3	3.8	20	14	71	12
		Polyester	23	20	12.6	2.5	8	8	32	9
		Open	15	0	7.8	-	13	-	26	-
23 March	Mulch	Clear	88	90	37.7	28.1	34	19	71	49
		Polyester	90	95	31.4	28.6	33	20	72	40
		Open	40	15	13.6	15.8	14	23	43	28
	None	Clear	40	70	21.1	17.4	14	16	59	26
		Polyester	28	65	3.5	6	18	12	16	14
		Open	10	5	1.9	0.2	9	11	4	0
13 April	Mulch	Clear	100	100	26.9	27.3	44	34	34	5
		Polyester	93	90	29.1	27.9	35	28	30	3
		Open	88	65	23.6	18.7	24	24	13	1
	None	Clear	95	93	26.5	18.9	31	33	10	1
		Polyester	73	45	18.5	10	25	12	3	0
		Open	45	38	11.2	7.7	23	29	0	0
LSD (5%)			*11*	*12*	*7.7*	*7.1*	*11*	*10*	*11*	*13*
Mean			*57*	*47*	*20.2*	*13.8*	*22*	*19*	*36*	*15*

Yield per hectare

1 Data are plot yields summed over eight harvests and averaged over replications. The experiment had a RCBD with a split-split-plot treatment structure: planting date was the whole-plot factor, chilling resistance level was the split-plot factor, and soil mulch and row cover were the split-split-plot factors.
2 Plant stand standardized by the best treatment stand (Black polyethylene mulch and clear polyethylene cover).
3 Total yield after eight harvests.
4 (Non-marketable yield × 100) / Total yield.
5 Percent of total yield after the first three of eight harvests.
6 Black polyethylene (mulch) vs. None.
7 Clear slitted polyethylene on wire hoops (clear) vs. Floating non-woven polyester (polyester) vs. Open.
8 'Wisconsin SMR 18', resistant to chilling at T<5°C.
9 'Poinsett 76', susceptible to chilling at T<5°C.

Table 2. Yield of chilling resistant and chilling susceptible cucumber cultivars planted under floating polyester covers with plastic mulch on 17 Mar. and 04 Apr. at Clinton, North Carolina 1988.

Planting date	Cultivar name	Chilling resistance	Total [4] Wt. (Mg)	No. (Th)	Marketable Wt. (Mg)	No. (Th)	Cull [2] (%)	Early [3] (%)
Pickling cucumbers								
17 March	Gy 14 A	S	43.0	808	33.3	624	22	65
	Calypso	S	40.3	715	33.6	595	16	64
	Castlepik	-	39.7	831	33.9	703	14	64
	Raleigh	-	38.8	758	32.2	625	17	57
	M 29	S	33.0	600	26.3	476	20	61
	Pixie	R	30.8	535	28.1	487	9	59
	M 28	S	30.4	536	22.8	404	25	66
	Wisconsin SMR 18	R	27.1	412	20.8	320	23	53
	M 21	R	24.9	468	20.6	384	17	53
	Sumter	S	24.8	383	21.6	335	13	41
	Wautoma	S	24.7	484	20.3	413	18	41
	Chipper	R	21.3	350	19.6	321	8	37
	H-19	R	16.2	294	14.8	266	9	17
	Albion	S	12.2	171	10.4	148	14	23
4 April	Castlepik	-	30.6	592	27.6	525	10	51
	Raleigh	-	29.8	633	26.0	550	13	58
	M 29	S	25.0	433	22.4	385	10	46
	Pixie	R	24.6	350	23.0	320	7	48
	Calypso	S	24.2	471	21.1	408	13	50
	Wisconsin SMR 18	R	22.7	366	18.5	295	18	45
	M 28	S	21.9	421	17.8	337	19	45
	Gy 14A	S	20.9	429	17.9	371	14	49
	Sumter	S	20.2	378	17.7	338	12	36
	M 21	R	18.2	340	16.9	308	7	28
	Chipper	R	15.8	267	14.4	242	9	29
	Wautoma	S	12.1	261	10.7	235	11	11
	H-19	R	11.6	259	10.9	242	6	5
	Albion	S	10.8	146	9.2	122	15	12
	LSD (5%)		*9.2*	*163*	*7.1*	*129*	*7*	*15*
Slicing cucumbers								
17 March	Supergreen	-	55.9	250	47.6	197	15	57
	Dasher II	S	52.2	217	46.9	191	10	59
	Centurion	S	51.4	219	41.7	164	19	51
	Sprint 440S	S	50.1	197	42.8	160	15	51
	Early Triumph	-	38.7	147	34.8	128	11	26
	Ashley	S	38.5	153	34.9	134	9	18
	Palomar	-	33.9	135	30.3	116	11	15
	Straight 8	S	33.5	133	25.4	89	28	45
	Pacer	-	30.0	127	27.1	111	9	26
	Mekty Green	-	29.8	62	20.1	41	33	12
	Marketmore 76	S	22.2	80	20.9	74	5	5
	Poinsett 76	S	18.3	71	17.9	69	2	17
	Tablegreen 65	S	11.5	43	10.1	36	12	4
	Lemon	-	4.3	44	4.2	44	2	3

[Continued on next page.]

Table 2 [continued].

4 April	Sprint 440S	S	48.4	199	37.7	143	12	65
	Supergreen	-	46.5	186	37.7	139	19	72
	Centurion	S	43.4	176	35.2	134	19	60
	Straight 8	S	41.2	156	31.7	108	24	57
	Dasher II	S	40.7	163	35.7	135	12	61
	Palomar	-	33.7	134	27.8	108	17	37
	Early Triumph	-	33.7	128	29.6	108	12	28
	Ashley	S	29.4	117	25.2	95	14	39
	Marketmore 76	S	26.5	107	22.4	86	15	18
	Pacer	-	26.3	110	22.9	92	13	28
	Poinsett 76	S	22.3	95	19.3	79	13	43
	Mekty Green	-	17.3	40	9.9	24	45	22
	Tablegreen 65	S	14.7	58	12.0	45	18	2
	Lemon	-	11.8	87	11.7	86	1	12
	LSD (5%)		*6.7*	*29*	*6.1*	*25*	*6*	*15*

1 Data are plot yields summed over six harvests and averaged over replications. The experiment had a RCBD with a split-split-plot treatment structure: planting date was the whole-plot factor, crop was the split-plot factor, and cultivar was the split-split-plot factor.

2 (Non-marketable yield × 100) / Total yield.

3 Percent of total yield after the first three of six harvests.

4 R = Resistant to chilling at T<5°C; S = Susceptible to chilling at T<5°C.

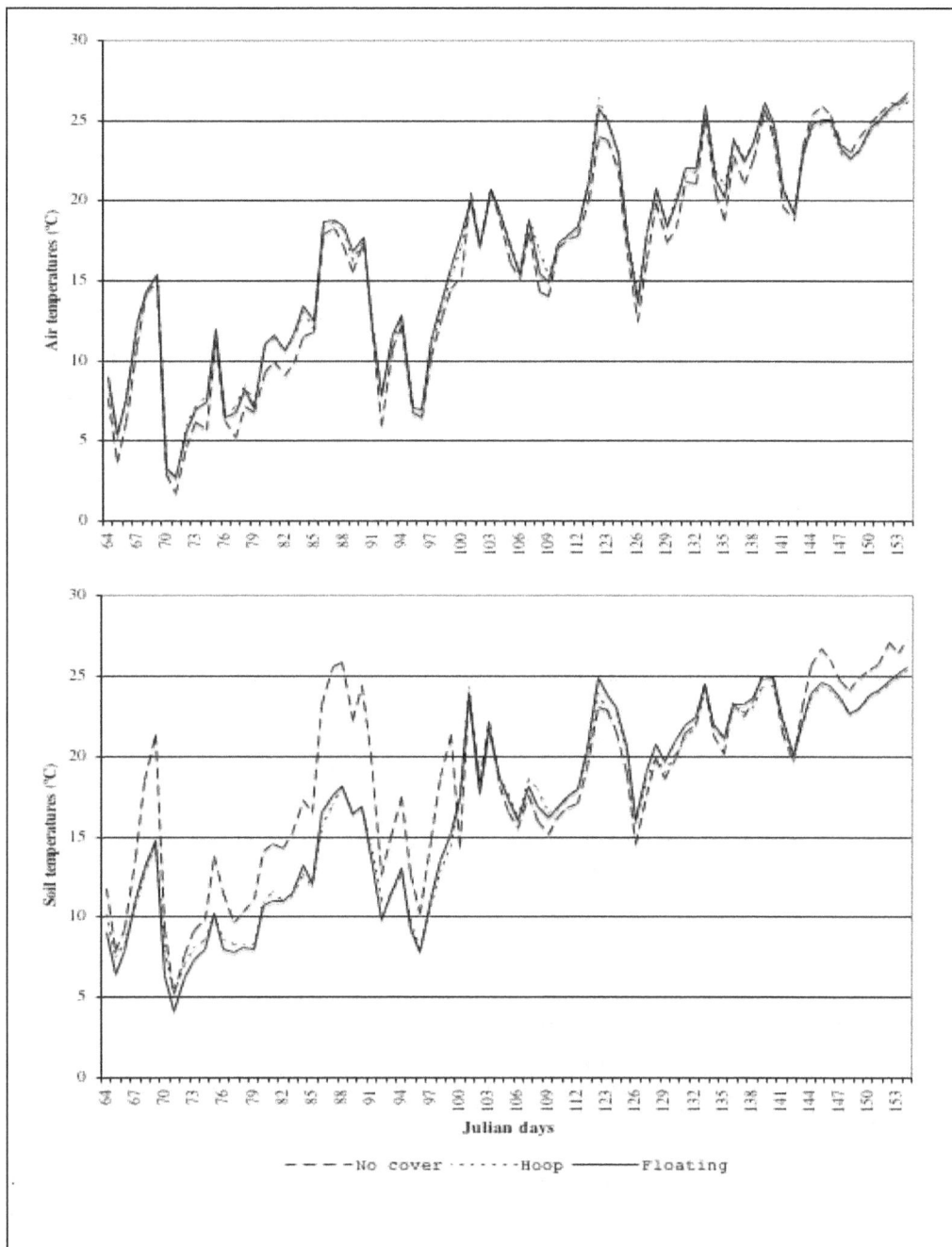

Figure 1. Average air and soil temperature of plors without mulch under different row covers during the trials of 14 pickling and 14 silcing cucumbers for early production at Clinton, North Carolina, 1987. The planning dates of 2 March, 23 March, and 13 April correspond to days 64, 85, and 106 of the Julian calender.

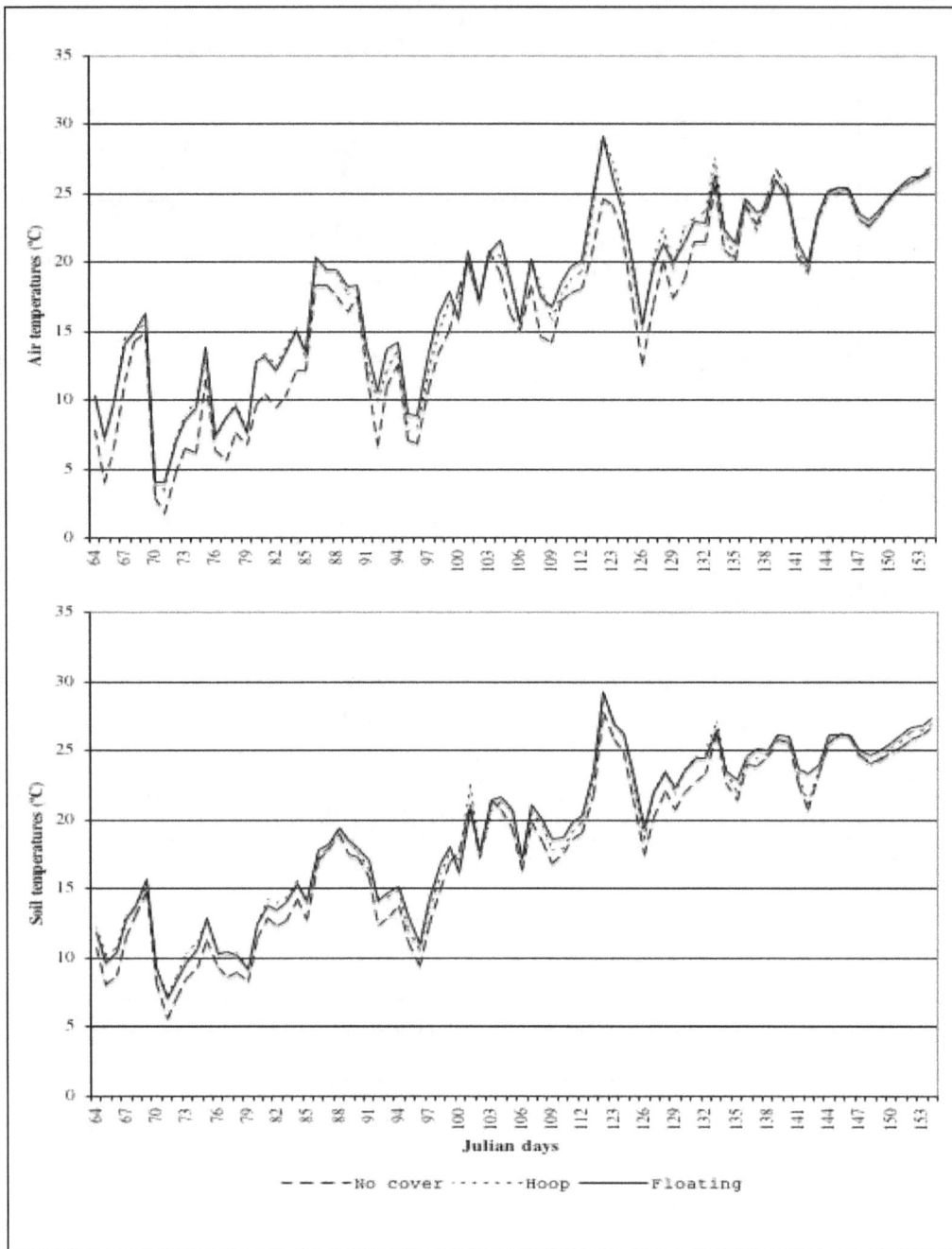

Figure 2. Average air and soil temperature of plots with black polyethylene mulch under different row covers during the trials of 14 pickling and 14 slicing cucumbers for early production at Clinton, North Carolina, 1987. The planting dates of 2 March, 23 March, and 13 April correspond to days 64m 85, and 106 of the Julian calendar.

Genetic Control of Downy Mildew Resistance in Cucumber - A Review

Adam D. Criswell, Adam D. Call and Todd C. Wehner
Department of Horticultural Science, North Carolina State University, Raleigh, NC 27695-7609

The oomycete pathogen *Pseudoperonospora cubensis* (Berk. and Curt.) Rostow. is a major foliar disease of cucumber (*Cucumis sativus* L.), especially in humid regions of the world (Palti and Cohen, 1980). Downy mildew was first described on cucumber by Berkeley and Curtis in 1868 and investigation into the genetic basis of resistance began in the early 20th century. Understanding the inheritance of downy mildew is fundamental to successful cucumber breeding programs. The objective of this article was to review the current knowledge of the inheritance of downy mildew resistance in cucumber.

Literature review

Early disease screening efforts at the Puerto Rico Agriculture Experiment Station focused on transferring resistance identified in a Chinese cultigen (Roque, 1937) into adapted varieties. The new Chinese cultivar introduced in 1933 was highly resistant, but had long and curved fruit that were not commercially usable. This cultivar was crossed with elite cultivars to combine the resistance with good horticultural traits. This eventually led to seven highly resistant lines having good characteristics. Of these, Puerto Rico selections 37, 39, and 40 were found to have resistance and good fruit quality, as well as yield superior to commercial cultivars used as checks.

Cochran (1937) used the Indian cultivar 'Bangalore' as a source of downy mildew resistance for crosses with popular slicing and pickling cultivars of the time. Cochran (1937) had some success with crosses to 'Bangalore' but did not determine the inheritance of resistance.

Jenkins (1946) used P.R. 37 as a resistant parent in studies of downy mildew in Minnesota as part of his Ph.D. dissertation research. He did not attempt to describe the inheritance of downy mildew resistance except to say that it was probably due to a number of factors. Part of his research involved a study of the correlation between physical traits and disease resistance. Of the traits observed (spine color, fruit color, fruit netting, spine texture, and growth habit) only growth habit appeared to have any relation to resistance. Jenkins suggested that determinate plants were more susceptible to

downy mildew than indeterminate plants. Barnes and Epps (1950) observed that even resistant plants became more susceptible to infection when fruit began to approach maturity. Determinate plants have concentrated fruit set, so that may explain their increased susceptibility to downy mildew.

'Palmetto', a cross between P.R. 40 and 'Cubit', was released in 1948 as a highly resistant slicing cucumber (Barnes, 1948). Resistance was attributed to two primary factors; high resistance to initial infection, exhibited by very few lesions, and limited sporulation resulting in decreased secondary infection. These resistance mechanisms were thought to be controlled by several genes. Limited acreage of 'Palmetto' was planted in 1948 and 1949. In those years, downy mildew was only found on 'Palmetto' when it was planted adjacent to susceptible cultivars, such as 'Marketer'. In 1950 and 1951, all 'Palmetto' fields inspected showed infection, regardless of proximity to 'Marketer' or other susceptible cultivars. In those years, the lesions were large and sporulated heavily, typical of lesions of susceptible cultivars (Epps and Barnes, 1952). Because it is unlikely that the change from 1948 to 1950 was due to a change in resistance, the change likely was in the pathogen population, either through mutation, selection in a mixed population, or migration of a race from a different region.

A new type of resistance was described by Barnes and Epps (1954) that was found in Cucumis sativus PI 197087 from India. PI 197087 was previously reported as having some resistance to downy mildew, as well as immunity to anthracnose (Barnes and Epps, 1952), The reaction of PI 197087 to downy mildew infection was characterized by small irregularly shaped, brown lesions, with a slight water-soaked appearance, becoming necrotic with sparse sporulation (Barnes and Epps, 1954), resembling a classic hypersensitive response (HR). Resistance from PI 197087 was used in the development of new cultivars, and the resistance in many current cultivars traces to PI 197087 (Wehner and Shetty, 1997). That resistance was effective for growers to produce cucumbers in warm humid regions of the U.S. without fungicides for over 40 years, but has not been as effective since a major epidemic in 2004.

Several studies have dealt with the inheritance of

downy mildew resistance in cucumber. Shimizu et al. (1963) reported that resistance in 'Aojihai' was controlled by three recessive genes (proposed s_1, s_2 and s_3) Pershin et al. (1988), using cultivar 'Sadao Rischu', determined resistance to be controlled by at least three major genes exhibiting partial dominance that were linked to at least three genes for powdery mildew resistance.

Van Vliet and Meysing (1974) concluded that downy mildew resistance from 'Poinsett', probably originating from PI 197087, was controlled by a single recessive gene that they named p. In addition, they proposed that the downy mildew gene was linked with the genes for powdery mildew resistance and for dull green fruit color (D). In a following study, Van Vliet and Meysing (1977) confirmed that the gene for hypocotyl resistance to powdery mildew was linked with or identical to the gene for resistance to downy mildew. They also concluded that the resistance found in 'Poinsett', 'Ashley', 'Taipei', 'Natsufushinari', PI 179676, and PI 234517 was controlled by the same gene. However, they stated that downy mildew resistance in 'Ashley' resulted from PI 197087, whereas resistance actually traced back to P.R. 40 (Barnes and Epps, 1956). This would explain why 'Poinsett' was reported to be more resistant than 'Ashley' and suggests that its resistance is due to a different gene.

Fanourakis and Simon (1987) reported agreement with Van Vliet and Meysing (1974) confirming that downy mildew resistance is controlled by a single recessive gene. They also reported loose linkage with powdery mildew resistance (pm) and compact plant (cp) genes. They reported a discrepancy in their results for one F_2 and one backcross family which did not fit the single-gene hypothesis. They attributed this to difficulty in rating resistance based on phenotypic expression at the cotyledon stage. No linkage with dull green fruit (D) was found, but there was deviation from expected results in one F_2 family.

El-Hafaz et al. (1990) report that the cultivars 'Palmetto' and 'Yomaki' were resistant in Egypt. They concluded that resistance was the result of an epistatic interaction between a dominant susceptible gene and a recessive resistance gene. Badr and Mohamed (1998) also determined that resistance was controlled by a pair of dominant and recessive interacting genes. Angelov (1994) reported that resistance in PI 197088 was due to two recessive genes and that 'Poinsett' resistance was inherited as a single recessive gene. PI 197088 was collected from the same region and at the same time as PI 197087.

Doruchowski and Lakowska-Ryk (1992) had evidence that downy mildew resistance was controlled by three recessive genes (dm-1, dm-2 and dm-3), where dm-3 and either dm-1 or dm-2 had to be homozygous recessive for maximum resistance. However, there was discrepancy in the F_2 results, which did not agree with their model. They argue that this resulted from testing too narrow a population. The three genes were included in the previous cucumber gene list (Pierce and Wehner, 1990), but should probably be removed as none of the genes were identified and no type lines are available to use in studies of separate genes. Petrov et al. (2000) claimed that the resistance in J-13, which was derived from Wisconsin 2843 (resistance originally from PI 197087 according to Peterson et al., 1985) was not inherited in a clear manner, but suggested it was due to one or two incompletely dominant genes.

PI 197088 was recently described as highly resistant to downy mildew in a large germplasm screening study and a multiple year re-evaluation of the most resistant and susceptible cultigens conducted at North Carolina State University (Criswell, 2008; Call, 2010). It appears that there are at least three genes for resistance to downy mildew in cucumber: one from the Chinese cultivar used in developing the PR lines, one from PI 197087, and one from PI 197088 (assuming that PI 197087 and PI 197088 share one resistance gene, dm-1). Inheritance studies are currently being conducted to determine if these genes are allelic.

Discussion

There are several proposed inheritance patterns for resistance to downy mildew. They range from three recessive genes (Doruchowski and Lakowska-Ryk, 1992; Shimizu et al., 1963) to three partially dominant genes (Pershin et al., 1988) to an interaction between dominant susceptible and recessive resistance genes (Badr and Mohamed, 1998; El-Hafaz et al., 1990) to one or two incompletely dominant genes (Petrov et al., 2000) to a single recessive gene (Angelov, 1994; Fanourakis and Simon, 1987; Van Vliet and Meysing, 1974; 1976). Conflicting results regarding the inheritance of downy mildew resistance in cucumber is likely due to four main factors.

First, the pathogen is highly variable and populations have not been well studied for the factors causing virulence (Lebeda and Urban, 2004). Multiple pathotypes and races have been identified (Lebeda and Widrlechner, 2003). In some cases, more than one pathotype in a geographical region has been identified (Lebeda and Urban, 2004). Different races have been reported (Angelov et al., 2000; Epps and Barnes, 1952; Hughes and Van Haltern, 1952; Shetty et al., 2002) and there may be different genes involved in resistance to different races. Call and Wehner (2010) noted a change in rank of resistant and moderate cultigens from screening studies before and after a change in the pathogen population. Cultigens highly resistant in 1988 and 1989 were only moderately resistant in studies conducted from 2005 to 2009. Those cultigens identified as highly resistant in the most recent studies were only moderately resistant in 1988 and 1999.

Second, differences in the environment, including temperature, humidity, rainfall and inoculum movement by wind all influence the severity of downy mildew infection (Cohen, 1977). Interactions between pathogen, host and environment are complex and not easily determined. Greenhouse tests are important for reducing environmental variability and should be conducted in addition to field tests. High variability in pathogen-host interactions due to environment can cause simply inherited traits to appear polygenic. This may be misleading and continuous variation with no clear segregation, even in homozygous inbred lines, can also indicate low heritability (Shaner, 1991). Horejsi et al. (2000) measured a low broad-sense heritability for downy mildew resistance, and noted large plant-to-plant variability in their study.

Third, different mechanisms of resistance have been studied (Angelov and Krasteva, 2000; Baines, 1991; Barnes and Epps, 1950; 1954; Palti and Cohen, 1980; Tarakanov et al., 1988). The previously mentioned inheritance studies used different mechanisms of resistance when evaluating plant response. Doruchowski and Lakowska-Ryk (1992) used necrotic lesions; Van Vliet and Meysing (1974; 1977) and El Hafaz et al. (1990) used sporulation intensity; Fanourakis and Simon (1987) used incidence of chlorotic and necrotic lesions on cotyledons; and Petrov et al. (2000) used chlorotic lesions for rating resistance. Other studies did not specify how resistance was measured. Different mechanisms of resistance may have different inheritance patterns.

Fourth, the original source of resistance varies over downy mildew inheritance studies. Some studies evaluated resistance sources from PI 197087 (India) while other studies evaluated resistance from P.R. 40 (China) and other germplasm sources. There are at least three genes for resistance to downy mildew, coming from P.R. 40, PI 197087, and PI 197088. Although, P.R. 40 is not available in the germplasm collection, cultivars tracing resistance to P.R. 40 are. Those include 'Ashley' and Ames 4833. The combination of the two different sources should provide either better resistance or more durable resistance. This combination can be found in PI 234517 (SC-50), which does have slightly higher resistance to downy mildew than 'Ashley' or PI 197087. However, the difference between PI 234517 and cultivars having resistance from PI 197087 alone was not significant (Wehner and Shetty, 1997). This is not surprising as

this resistance was overcome in 1950 in the cultivar 'Palmetto'.

Conclusions

There is evidence that PI 197087 currently in the germplasm collection has lost resistance to downy mildew. Barnes and Epps (1954) described the phenotypic response of PI 197087 as having small necrotic lesions with no chlorosis. Yet, Van Vliet and Meysing (1974) reported resistance in 'Poinsett' as necrotic spots on cotyledons with small chlorotic spots on true leaves. Shetty et al. (2002) described resistance as small necrotic or chlorotic spots, Chipper and Poinsett 76 both exhibited these symptoms. Wehner and Shetty (1997) state that PI 197087 showed only intermediate resistance, with chlorotic and necrotic spots. On the other hand, 'Chipper', Gy 4, M 21, 'Poinsett 76', 'Pixie', and 'Polaris' were all more resistant than PI 197087, even though their resistance was derived from PI 197087. Angelov (1994) reported that 'Poinsett' had only moderate resistance, with chlorotic lesions present on the leaves. Petrov et al. (2000) reported that resistance in WI 2843 (from PI 197087) was expressed as small chlorotic lesions. However, in their study 'Poinsett' showed small necrotic lesions. The loss of the typical resistance response phenotype in PI 197087 may be due to a combination of downy mildew overcoming the hypersensitive response described by Barnes and Epps (1954) along with genetic drift in the PI collection.

There probably are multiple genes for downy mildew resistance, each with a distinct inheritance pattern. However, the cucumber gene list should be corrected by removal of *dm-1*, *dm-2* and *dm-3* genes, while keeping the *dm* gene from PI 197087. Additional genes, for example from P.R. 40, could be included when their effects have been isolated and type lines identified.

Literature Cited

Angelov, D. 1994. Inheritance of resistance to downy mildew, *Pseudoperonospora cubensis* (Berk. & Curt.) Rostow. Rep. 2nd Natl. Symp. Plant Immunity (Plovdiv) 3:99-105.

Angelov, D., P. Georgiev, and L. Krasteva. 2000. Two races of *Pseudoperonospora cubensis* on cucumbers in Bulgaria. Proc. Cucurbitaceae 2000. (N Katzir and H. S. Paris eds.), pp. 81-83. ISHS Press, Ma'ale Ha Hamisha, Israel. 509 p.

Angelov, D. and L. Krasteva. 2000. Selecting downy mildew-resistant short-fruited cucumbers. Proc. Cucurbitaceae 2000. N. Katzir and H.S. Paris (eds.), pp. 135-137. ISHS Press, Ma'ale Ha Hamisha, Israel. 509 p.

Badr, L.A.A. and F.G. Mohamed. 1998. Inheritance and nature of resistance to downy mildew disease in cucumber (*Cucumis sativus* L.). Annals of Agriculture Science, Moshtohor (abstract only) 36(4):2517-2544.

Bains, S.B. 1990. Classification of cucurbit downy mildew lesions into distinct categories. Indian J. Mycology Plant Pathology 21(3)269-272.

Barnes, W.C. 1948. The performance of Palmetto, a new downy mildew resistant cucumber variety. Proc. Amer. Soc. Hort. Sci. 51:437-441.

Barnes, W.C. and W.M. Epps. 1950. Some factors related to the expression of resistance of cucumbers to downy mildew. Proc. Amer. Soc. Hort. Sci. 56:377-380.

Barnes, W.C. and W.M. Epps. 1954. An unreported type of resistance to cucumber downy mildew. Plant Disease Reporter 38(9):620.

Barnes, W.C. and W.M. Epps. 1956. Powdery mildew resistance in South Carolina cucumbers. Plant Disease Reporter 40(12):1093.

Berkeley, M.S. and A. Curtis. 1868. *Peronospora cubensis*. J. Linn. Soc. Bot. 10:363.

Call, A.D. 2010. Studies on resistance to downy mildew in cucumber (Cucumis sativus L.) Caused by Pseudoperonospora cubensis. M.S. Thesis, North Carolina State Univ., Raleigh.

Call, A.D. and T.C. Wehner. 2010. Search for resistance to the new race of downy mildew in cucumber, p. 112-115. Cucurbitaceae 2010 Proceeding. ASHS Press, Alexandria, VA

Cochran, F.D. 1937. Breeding cucumbers for resistance to downy mildew. Proc. Amer. Soc. Hort. Sci. 35:541-543.

Cohen, Y. 1977. The combined effects of temperature, leaf wetness, and inoculum concentration on infection of cucumbers with *Pseudoperonospora cubensis*. Canadian J. Botany 55:1478-1487.

Criswell, A.D. 2008. Screening for downy mildew resistance in cucumber. M. S. Thesis, North Carolina State Univ., Raleigh.

Doruchowski, R.W. and E. Lakowska-Ryk. 1992. Inheritance of resistance to downy mildew (Pseudoperonospora cubensis Berk & Curt) in Cucumis sativus. Proc. 5th EUCARPIA Symp. (R. W. Doruchowski, E. Kozik, and K. Niemirowicz-Szczytt, eds.), pp. 132-138, 27-31 July, Warsaw, Poland. Published by Res. Inst. Veg. Crops, and Warsaw Univ. Agric., Warsaw, Poland.

El-Hafaz, A., B. El-Din, H.H. El-Doweny, and M.M.W. Awad. 1990. Inheritance of downy mildew resistance and its nature of resistance in cucumber. Annals of Agricultural Science, Moshtohor. 28(3):1681-1697.

Epps, W.M. and W.C. Barnes. 1952. The increased susceptibility of the Palmetto cucumber to downy mildew in South Carolina. Plant Disease Reporter 36(1):14-15.

Fanourakis, N.E. and P.W. Simon. 1987. Analysis of genetic linkage in the cucumber. J. Heredity 78:238-242.

Horejsi, T., J.E. Staub, and C. Thomas. 2000. Linkage of random amplified polymorphic DNA markers to downy mildew resistance in cucumber (*Cucumis sativus* L.). Euphytica 115:105-113.

Hughes, M.B. and F. Van Haltern. 1952. Two biological forms of *Pseudoperonospora cubensis*. Plant Disease Reporter 36(9):365-367.

Jenkins, J.M. Jr. 1946. Studies on the inheritance of downy mildew resistance and of other characters in cucumbers. J. Heredity 37:267-271.

Lebeda, A. and M.P. Widrlechner. 2003. A set of Cucurbitaceae taxa for differentiation of *Pseudoperonospora cubensis* pathotypes. J. Plant Diseases and Protection 110: 337-349.

Lebeda, A. and J. Urban. 2004. Disease impact and pathogenicity variation in Czech populations of *Pseudoperonospora cubensis*, pp. 267-273 In: A. Lebeda and H.S. Paris (eds). Progress in Cucurbit genetics and breeding research. Proc. Cucurbitaceae 2004, 8th EUCARPIA Meeting on Cucurbit Genetics and Breeding. Palacky University in Olomouc, Olomouc, Czech Republic.

Palti, J. and Y. Cohen. 1980. Downy mildew of cucurbits (*Pseudoperonospora cubensis*): The fungus and its hosts, distribution, epidemiology and control. Phytoparasitica 8:109-147.

Pershin, A.F., N.I. Medvedeva, and A.V. Medvedev. 1988. Quantitative approach to studying the genetics of disease resistance. IV. Interaction of the genetic systems for resistance to powdery and downy mildews in cucumber. Geneticka, USSR. (abstract only) 24(3):484-493.

Peterson, C.E., J.E. Staub, M. Palmer, and L. Crubaugh. 1985. Wisconsin 2843, a multiple disease resistant cucumber population. HortScience 20(2):309-310.

Petrov, L., K. Boodert, L. Sheck, A. Baider, E. Rubin, Y. Cohen, N. Katzir, and H.S. Paris. 2000. Resistance to downy mildew, *Pseudoperonospora cubensis*, in cucumbers. Acta Horticulturae 510:203-209.

Pierce, L.K. and T.C. Wehner. 1990. Review of genes and linkage groups in cucumber. HortScience 25:605-615.

Roque, A. 1937. Annual report of the agricultural experiment station of Puerto Rico fiscal year 1936-1937. pp. 45-46.

Shaner, G. 1991. Genetic resistance for control of plant disease. In: David Pimentel (ed.) CRC handbook of pest management in Agriculture Vol. 1. pp. 495-540.

Shetty, N.V., T.C. Wehner, C.E. Thomas, R.W. Doruchowski, and V.K.P. Shetty. 2002. Evidence for downy mildew races in cucumber tested in Asia, Europe, and North America. Scientia Horticulturae 94(3-4):231-239.

Shimizu, S., K. Kanazawa, A. Kato, Y. Yokota, and T. Koyama. 1963. Studies on the breeding of cucumber for the resistance to downy mildew and other fruit characters. Engei Shikenjo ho koku (abstract only) 2:65-81.

Tarakanov, G.I., A.V. Borisov, and S.O. Gerasimov. 1988. Methodology of breeding cucumber for resistance to downy mildew. Selektsiya, semenovodstvo I sortovaya tekhnologiya proizvodstva ovoshchei (abstract only) pp.13-17.

Van Vliet, G.J.A. and W.D. Meysing. 1974. Inheritance of resistance to *Pseudoperonospora cubensis* Rost. in cucumber (*Cucumis sativus* L.). Euphytica 23:251-255.

Van Vliet, G.J.A. and W.D. Meysing. 1977. Relation in the inheritance of resistance to *Pseudoperonospora cubensis* Rost and *Sphaerotheca fuliginea* Poll. in cucumber (*Cucumis sativus* L.). Euphytica 26:793-796.

Wehner, T.C. and N.V. Shetty. 1997. Downy mildew resistance of the cucumber germplasm collection in North Carolina field tests. Crop Science 37: 1331-1340.

Antagonistic Actinomycete XN-1 from Phyllosphere Microorganisms of Cucumber to Control *Corynespora cassiicola*

Minggang Wang and Qing Ma*

National Key Laboratory of Crops Stress Biology in Arid Areas and College of Plant Protection, Northwest A&F University, Yangling 712100, P. R. China

Strain XN-1 isolated from the cucumber phyllosphere was tested for potential application as an antagonistic actinomycete against Corynespora cassiicola (Berk.& Curt.)Wei. The fungistatic activity of this strain was determined by inoculation of C. cassiicola on PDA plates, spore germination inhibition in culture filtrate and control tests on cucumber leaves. The spore germination and hyphal growth inhibition rate of the XN-1 fermentation filtrate against C. cassiicola can reach 96.50% and 51.17%, respectively, as well as the 63.54% of control effect on cucumber leaves. The stability of XN-1 fermentation filtrates was studied under the treatment of pH gradient, temperature gradient, and exposure to UV. The results showed the filtrates remained relatively stable in the range of those gradients. This experiment indicates that this exogenous actinomycete XN-1 has the potential to act as an antagonistic agent in controlling the occurrence and development of cucumber target leaf spot in the greenhouse. This also confirms that phyllosphere microorganisms play an important role in combating the infection of pathogens and have a promising future in developing the biocontrol products and methods.

Corynespora cassiicola is regarded as a widespread pathogen, associated with more than 70 different host plants in tropical and subtropical countries (Pollack and Stevenson 1973; Onesirosan et al. 1974). In recent years, this fungus has resulted in a great loss on cucumber cultivated in plastic greenhouses in spite of treatment with various fungicides. Target leaf spot caused by *C. cassiicola* has been an important disease of cucumber since the end of the 1970s in Japan (Hasama 1990). Recently, the occurrence of this disease was also found in Korea (Kwon et al. 2003). While, in China, cucumber target leaf spot was reported as a new plant disease (Zou et al. 2002). It is well known that cucumber target leaf spot is difficult to manage in the greenhouse because of the optimum environmental conditions that are often conducive to disease development (Menzies and Belanger 1996). Moreover, the pathogen has become readily resistant to fungicides because of prolonged and continuous use. Therefore, the search for alternatives to chemical control of plant disease, especially biological control, has gained momentum in recent years (Compant et al.2005). The control of pathogens by biological agents has not been practiced on a large scale, but many experiments have shown the potential for further development as a promising approach in responding to disease control and consumers' health concerns. In recent years, some special phyllosphere microbes have been found to have great effects on the location of pathogens (Zhao et al. 2000).

Among microbes, the actinomycetes are important producers of bioactive compounds and represent a high proportion of the microbial biomass. Some actimomycetes secrete herbicidal compounds (Tanaka and Omura 1993) or fix atmospheric nitrogen; others can protect plants against fungal infections. The antagonistic impact of actinomycetes on pathogenic fungi is known and several species have been used as biological control agents (Jones and Samac 1996; Bressan 2003). , Most species which are isolated from the soil cannot be used as a biological agent directly. In this experiment, we sought to isolate microbes from the leaves so that some new antagonistic actinomycete isolates that exhibit a highly inhibitory effect against plant pathogens can be found.

Materials and methods

Sampling

The cucumber plants were collected using Zhao xinhua and Chen weiliang method (2000) from Gardenspot of Northwest University of Agriculture and Forest (Yangling, ShaanXi province, China). The leaf samples picked were placed in sterile polythylene bags, closed tightly and stored in the refrigerator at 4°C until use. The pathogen fungi were supplied by the Plant Pathology laboratory.

Isolation of phyllosphere actinomycetes

The picked leaves were placed into 200 ml sterile distilled water, and then shaken for 30min at room temperature. The suspension obtained was diluted by a

---

*Corresponding author: maqing@nwsuaf.edu.cn

factor of 10^{-1}, 10^{-2}, 10^{-3}; 0.1 ml of each dilution was placed in sterile petri dishes containing Actinomycetes Isolation Agar (AIA, Olson 1968) (5% glycerol, 0.2% sodium caseinate, 0.01% L-asparagine, 0.4% sodium propionate, 0.05% K_2HPO_4, 0.0001% $FeSO_4$ and 1.5% agar Difco). which was supplemented with 25% lactic acid to inhibit the development of bacteria without affecting the growth of fungi and actinomycetes (Naureen et al. 2009). Petri dishes were incubated for 3 days at 28°C. The isolates were purified and transferred to AIA media and stored at 4°C.

Antifungal assays

The antifungal activity of the isolates was determined by the plate diffusion method (Barakate et al.2002) against C. cassiicola. Isolates were grown on Bennett medium for 14 days and 5-mm diameter agar disks containing actionmycete colonial mass were prepared using sterile borers. Disks were then transferred aseptically to PDA plates in which fungal mycelial disk (5 mm diameter) was placed in the centre. Plates were incubated at 25°C. Inhibition zones were determined after 3 days of incubation. The isolates that showed an inhibition zone greater than 8 mm were considered as active ones, which were selected to be further studied.

Culture filtrate preparation of actinomycetes

Liquid cultures were grown in 250 ml flasks containing 100 ml of nutrient broth (NB) which had previously been added two sterile bacteriological loops of the culture. Flasks were incubated on a shaker at 180rpm at 28°C for 72 h. The culture was transferred to 50 ml centrifugation tubes, and centrifuged at 10000 rpm/min for 20 minutes, and then the supernatant of centrifuged cultures of antagonist was filtered through a 0.22 mm polycarbonate membrane filter to be culture filtrates.

Antagonism in vitro

In order to test the antagonism of culture filtrates against C. cassiicola, 5-mm diameter disks of agar from three-day-old C. cassiicola grown on PDA were transferred to the centre of plates containing 1 ml filtrates mixed into 10 ml molten PDA. Three days later, the diameter of pathogen colony was recorded. Results were expressed as mean % inhibition of the growth of C. cassiicola in the absence of the actinomycete. Percent inhibition was calculated using the following formula: % inhibition= (1-[fungal growth/control growth])×100.

The effect of liquid cultures on spore germination was also estimated in this experiment. The primary filtrate and the C. cassiicola spores were made into spores suspension of $5×10^4$ spores ml^{-1}. One-hundred-microliter (100ml) of the suspension was placed on a hollow-groundslide kept in a sterile Petri dish of which the bottom was covered by a filter paper saturated with SDW. The Petri dishes were then stored at 25°C and observations were made to check the percent of spore germination and inhibition after 12h. Each treatment had three replicates. The percent of germination was calculated using the formula: % germination= (number of germinated spores/number of checked spores)×100. % inhibition= (percent of germination of control – percent of germination of treatment/ percent of germination of control)×100. The experiment was repeated twice.

Inoculation plant assays

Cucumber plantlets grown for 6 weeks were used for the inoculation. Uniform plantlets (n=24) were selected for each treatment in this experiment (Ait Barka et al. 2000). The pathogen C. cassiicola was incubated under 0.2 mmol/m²/s white fluorescent light for 5d. The spore suspension was made and adjusted to 10^4 spores/ ml by plate counting, and then evenly sprayed on the surface of the cucumber leaves after the treatment of XN-1 filtrate. Control plants were treated only with sterile distilled water. Both treatments were kept moist in the next two days before the plants were transferred out from moisturizing device. Another three days later, the target leaf spot symptoms were evaluated.

Results

Evaluation of antagonistic activity of XN-1 , the effect of filtrate on C. cassiicola in vitro and on cucumber leaves

On PDA plates, XN-1 inhibited the growth of C. cassiicola by the presence of a transparent inhibition zone between the two organisms. There was a significant zone of inhibition around the actinomycete inoculum when the fungus was grown with the actinomycetes isolates on the same PDA plate (Fig. 1A). The hyphal growth inhibition rate was 78.34% under the treatment of XN-1 filtrate, which showed a potential for further study of this isolate (Fig. 1B). In the inoculation experiments, the cucumber leaves treated with culture filtrate of XN-1 had a few small lesions, whereas the control leaves exhibited typical symptoms of target leaf spot (Fig. 1C). The control rate of 63.54% indicated that the XN-1 markedly restricted the spread of the pathogen on cucumber leaves.

The inhibition of isolate XN-1 against the spore germination of C. cassiicola

In the results of detecting the inhibitory effects of XN-1 filtrate on the spore germination, the inhibition rate of primary filtrate could reach 96.50%. The formation of germtubes could not successfully happen from either or both poles of the spores, and in some cases, a spherically intumescent germtube could be observed (Fig. 2B), which might be deformed and fail to grow further.

The stability of culture filtrate

The culture filtrate of the strain XN-1 was stable and relatively high for the whole range of temperatures, although the inhibition rate decreased slightly with the increase of temperature (Fig. 3).

Culture filtrate of the strain was also mainly stable against the pathogen at pH values between 3.0 and 11.0 (Fig. 4). The effects of pH on inhibition rate of pathogen decreased with the increase of alkalinity or acidity, but the rate was still above 50% except for extreme alkalinity.

The effect of culture filtrate of strain XN-1 on *C. cassiicola* decreased as the length of exposure to ultraviolet radiation was extended (Fig. 5). But the inhibition rate maintained a relatively high level at over 40%, even being exposed for 45 min .

Discussion

In this study, XN-1 was isolated from the cucumber phyllosphere and tested in terms of spore germination, interactive reaction and inoculation assay to assess its future as a promising biocontrol agent. Being different from microorganisms selected from the soil, phyllosphere organisms can have direct effects on the target disease.

In the antifungal assays, some other actinomycetes were also found to exhibit marked antagonistic effects against *C. cassiicola*. However, the bioactive compounds of these agents did not show any antagonism against the pathogen, which might be attributed to the accumulation of the antibiotics influenced by type of the fermentation broth, primary pH value, temperature of shaker, the length of fermentation, and so on. Therefore, the fermentation conditions of these agents should be studied for a better application.

In addition, the identification of XN-1 and its colonization on the cucumber leaves deserve further studies for a better understanding of exact plant defense mechanisms, especially in the field conditions. Anyway, this study still elucidated a new path for controlling the cucumber target leaf spot.

Acknowledgments

This project was supported by the National Natural Science Foundation of China (grant no. 30771398), the 111 Project from Ministry of Education of China (B07049), the National Science & Technology Pillar Program in the 11th Five-year Plan (2006BAD07B02)

Referrences

1. Ait B. E., Belarbi A., Hachet C., Nowark J., Audran J. C. 2000. Enhancement of in vitro growth and resistance to gray mould of *Vitis vinifera* L. co-cultured with plant growth-promoting rhizobacteria. FEMS Microbiol Lett, 186: 91-95

2. Barakate M., Ouhdouch Y., Oufdou K., Beaulieu C. 2002. Characterization of rhizospheric soil streptomycetes from Maroccan habitats and their antimicrobial activities. World J Microbiol Biotechnol 18: 49-54

3. Bressan W. 2003. Biological control of maize seed pathogenic fungi by use of actinomycetes. BioControl, 48: 233-240

4. Compant S., Duffy B., Nowak J., Christophe C and Barka E A. 2005. Use of plant growth-promoting bacteria for biocontrol of plant diseases: principles, mechanisms of action, and future prospects. Applied and Environmental Microbiology, 71 (9): 4951-4959

5. Hasama W. 1990. Present status of occurrence changes, research and control of *Corynespora* cucumber target leaf spot. Plant Prot, 44: 224-228

6. Jones C. R. and Samac D. A. 1996. Biological control of fungi causing alfalfa seedling damping-off with a disease-suppressive strain of *Streptomyces*. Biol. Control, 7: 196-204

7. Kwon J. H., Jee H. J., and Park C. S. 2005. *Corynespora* leaf spot of balsm pear caused by *Corynespora cassiicola* in Korea. Plant Pathol J, 21 (2): 164-166

8. Menzies J. G., Belanger R. R. 1996. Recent advances in cultural management of diseases of greenhouse crops. Can. J Plant Pathol, 18: 186-193

9. Olson E. H. 1968. Actinomycetes Isolation Agar. In: Difco: supplementary literature. Difco Lab, Detroit, (Michi)

10. Onesirosan P. T., Arny D. C., and Durbin R. D. 1974. Host specificity of nigerian and north American isolates of *Corynespora cassiicola*. Phytopathology 64: 1364-1367

11. Pollack F. G. and Stevenson J. A. 1973. A fungal pathogen of *Broussonetia papyrifera* collected by George Washington Carver. Plant Dis Rep, 57: 296-297

12. Zou Q. D., Fu J. F., Zhu Y., and Pang D. C. 2002. Identification of pathogen from cucumber target leaf spot and research on biological characteristics. Journal of Shenyang Agricultural University, 33 (4): 258-261

13. Tanaka Y. and Omura S. 1993. Agroactive compounds of microbial origin. Annual Review of Microbiology, 47: 57-87

14. Zhao X. H., Chen W. L., and Li D. B. 2000. Selection of antagonistic bacteria to *Xanthomonas oryzae* pv. *oryzae* and identification of phyllosphere microbes on rice leaf. Chinese J. Rice Sci., 14 (3): 161-164

Fig. 1 A. The antagonism between XN-1 and *C. cassiicola* B. The inhibition effects of XN-1 filtrate on the hyphal growth of *C. cassiicola* C. Cucumber leaves of treatment by XN-1 fermentation filtrates and the control.

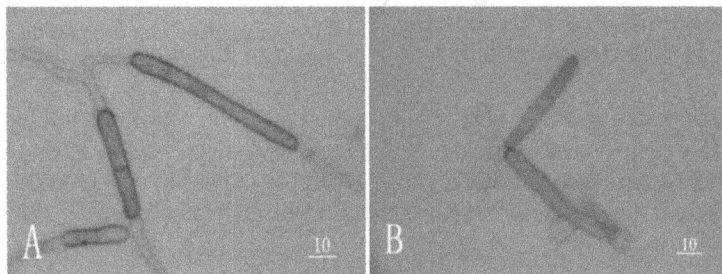

Fig. 2 A. The normal spore germination of *C. cassiicola* within water drop B. The inhibition of XN-1 filtrate against spore germination of *C. cassiicola* Note: The bar represents 10 μm

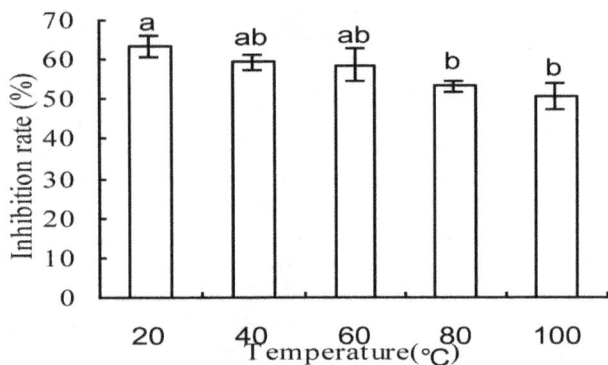

Fig. 3 The effects of temperature on inhibition rate of culture filtrates against *C. cassiicola.*.Inhibition rates are the mean of three trials. Values followed by the different letters are statistically different at P = 0.05 according to Duncan's multiple range test.

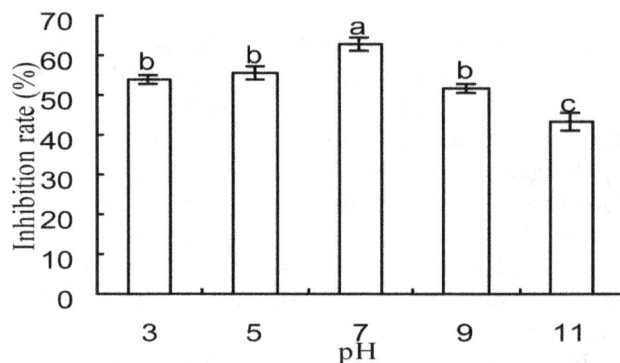

Fig. 4 The effects of pH on inhibition rate of culture filtrates against *C. cassiicola.*. Bars represent standard deviations of the means. Values followed by the same letters are not statistically different at P =0.05 according to Duncan's multiple range test.

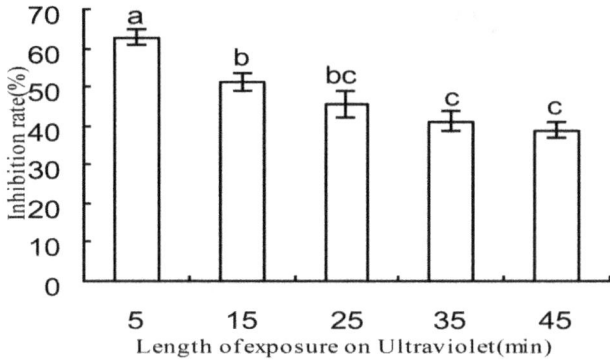

Fig. 5 The effects of exposure on Ultraviolate on inhibition rate of culture filtrates against *C. cassiicola*. Bars represent standard deviations of the means. Values followed by the same letters are not statistically different at P =0.05 according to Duncan's multiple range tests.

Powdery Mildew Resistance of Cucurbits at Different Locations

R. W. Robinson

Horticulture Dept., Cornell University, Geneva, NY 14456

A sabbatical leave in 1968/69 provided an opportunity for me to visit research institutes in five different countries and test cucurbits for powdery mildew resistance at each location. Sixteen accessions of six cucurbit species were inoculated with powdery mildew spores in greenhouses at Geneva, NY; La Jolla, CA; Wageningen, Holland; Fischenich, Germany; Budapest, Hungary; and Littlehampton, England. Results of this survey of more than 40 years ago were mentioned in correspondence by G. W. Bohn that was recorded by McCreight (9), but they have not been previously published.

Cucumis melo: 'Hearts of Gold' melon was susceptible, whereas 'Perlita', 'Dulce', 'Campo', 'Jacumba', 'Big River', and 'PMR 5' were resistant at each location. 'PMR 45' and 'Gulfstream' melons were susceptible at La Jolla, CA but resistant at all other locations. 'PMR 45' is resistant to race 1 but susceptible to race 2 of cucurbit powdery mildew (7). Race 2 evidently occurred at La Jolla, CA but not at the other locations in this survey since 'PMR 45' was susceptible only at La Jolla.

The resistance of 'PMR 45' to race 1 of cucurbit powdery mildew is conditioned by a single dominant gene, Pm-1 (8). 'Gulfstream' had the same reaction to powdery mildew at each location as 'PMR 45' and presumably has the same resistance gene. The pedigree of 'Gulfstream' supports this conclusion since 'PMR 45' was one of its parents (1).

Although 'PMR 45' was resistant to powdery mildew at Geneva NY when this survey was made, it was susceptible there in one out of nine years in subsequent tests. Thus, both race 1 and race 2 have occurred there but race 1 has been predominant.

'PMR 5', which was resistant at La Jolla and all other locations tested, is resistant to races 1 and 2 of powdery mildew and has gene Pm-2 and modifiers for race 2 resistance (2). It is considered to also have the Pm-1 gene for race 1 resistance (6). 'Big River' had the same reaction to powdery mildew at each location as 'PMR 5' and presumably has the same resistance genes. 'Big River' is a selection of 'PMR 6 (1), which has similar parentage as 'PMR 5' (6). Cohen and Eyal (3) reported evidence that 'PMR 6' as well as 'PMR 5' has the Pm-2 gene; they found that 'PMR 5' and 'PMR 6' were both resistant to race 1 and 2 in Israel.

'Campo', 'Jacamba', 'Dulce', and 'Perlita' were also resistant at each location and hence have resistance to both race 1 and 2. Race 1 resistance was associated with race 2 resistance in this survey and also in the tests of Cohen and Eyal (3); they determined that 16 of the 43 cultigens they tested were resistant to race 2 and race 1. Harwood and Markarian (6) noted this association of race 1 and 2 resistance and concluded that 'PMR 5' has the Pm-1 gene for race 1 resistance and Pm-2 for resistance to race 2. If this is the case and Pm-1 and Pm-2 are independent genes, then some cultivars bred for race 2 resistance without selection for race 1 resistance might lack the Pm-1 gene and be susceptible to race 1. None of the race 2 resistant cultivars in these tests, however, were susceptible to race 1. Combined resistance to both races could result if Pm-1 and Pm-2 are closely linked or if Pm-2 provides resistance for both races.

When this survey was made powdery mildew of melon and other cucurbits was considered to be caused by *Erysiphe cichoracearum*, but subsequently *Sphaerotheca humuli* var. *fuligenia* (since revised to *Podosphaera xanthii*), was identified as the causal agent (9). Correspondence of G. W. Bohn indicates that race 2 resistant melons were resistant at nearly all sites tested in various parts of the world and he concluded that *Sphaerotheca humilis* var. *fuligenia* was the prevalent species causing powdery mildew of cucurbits (9). Results of this survey corroborate that conclusion.

Cucumis sativus: C7-63 (a cucumber breeding line from C. E. Peterson) and 'Yamaki' were resistant to both race 1 and 2, since they were resistant at La Jolla CA where race 2 was present and at the other locations having race 1. 'Lemon' was susceptible at each location.

Cucumis anguria: *C. anguria* was resistant but not immune to powdery mildew. Only a trace of powdery mildew occurred on this species at Littlehampton, England and no evidence of powdery mildew was observed on *C. anguria* at the other locations. *C. anguria*, therefore, is resistant to both races.

Cucurbita: *C. lundelliana* was resistant and *C. pepo* cv. Scallop was susceptible at each location tested. *C.*

lundelliana has a single dominant gene for resistance to powdery mildew (10). *C. okeechobeensis* subsp. *martinezii*, which has the same resistance gene (4), was used by H. M. Munger to breed germplasm that is the basis for the powdery mildew resistance of squash and pumpkin varieties developed by seed companies. These cultivars, therefore, should be similar to *C. lundelliana* in being resistant to both race 1 and race 2 of powdery mildew.

Citrullus lanatus: Powdery mildew was not considered an important disease of watermelon when this survey was made, but recently it has become a serious problem in some locations (5). 'Klondike R7' watermelon was resistant at each location in this survey and therefore was resistant to both race 1 and 2.

Acknowledgements

This survey of powdery mildew resistance was made possible by the gracious cooperation of G. W. Bohn and T. W. Whitaker (U. S. Dept. of Agriculture Horticultural Field Station at La Jolla, CA), G. E. G. Crüger (Biologische Bundesanstat für Land- und Forstwirtschaft, Fischenich, Germany), A. Andrasfalvy (Vegetable Crops Research Institute, Budapest, Hungary), P. E. Grimbly and L. A. Darby (Glasshouse Crops Research Institute, Littlehampton, England), and E. Drijfhout and E. Kooistra (Institute for Horticultural Plant Breeding, Wageningen, Holland).

Literature Cited:

1. Bohn, G. W., C. F. Andrus, and R. T. Correa. 1969. Cooperative muskmelon breeding program in Texas, 1955-67: new rating scales and index selection facilitate development of disease-resistant cultivars adapted to different geographical areas. USDA Tech. Bull. 1405.

2. Bohn, G. W. and T. W. Whitaker. 1964. Genetics of resistance to powdery mildew race 2 in muskmelon. Phytopathology 54: 587-591.

3. Cohen,Yigal and Helena Eyal. 1988. Reaction of muskmelon genotypes to races 1 and 2 of *Sphaerotheca fuligenia*. Cucurbit Genetics Coop. Rpt. 11: 47-49.

4. Contin, M. E. 1978. Interspecific transfer of powdery mildew resistance in the genus *Cucurbita*. PhD thesis, Cornell Univ., Ithaca NY.

5. Davis, A. R., B. D. Bruton, S. D. Pair, and C. E. Thomas. 2001. Powdery mildew: an emerging disease of watermelon in the United States. Cucurbit Genetics Coop. Rpt. 24: 42-48.

6. Harwood. R. R. and D. Markarian. 1968. A genetic survey of resistance to powdery mildew in muskmelon. J. Hered. 59: 213-217.

7. Jagger, I. C., T. W. Whitaker, and D. R. Porter. 1938. A new biologic form of powdery mildew on muskmelons in the Imperial Valley of California. Plant Dis. Reptr. 51: 1079-1080.

8. Jagger, I. C., T. W. Whitaker, and D. R. Porter. 1938. Inheritance in Cucumis melo of resistance to powdery mildew (Erysiphe cichoracerarum). Phytopathology 28: 671.

9. McCreight, James D. 2004. Notes on the change of the causal species of cucurbit powdery mildew in the U. S. Cucurbit Genetics Coop. Rpt. 27: 8-23.

10. Rhodes, A. N. 1964. Inheritance of powdery mildew resistance in the genus *Cucurbita*. Plant Dis. Reptr. 48: 54-55.

Podosphaera xanthii but not *Golovinomyces cichoracearum* Infects Cucurbits in a Greenhouse at Salinas, California

Cosme Bojorques Ramos
Research Center for Food and Development (CIAD, A. C.), Department of Phytopathology, km 5.5 Culiacan-Eldorado Road, Culiacan, Sinaloa, Mexico, and University of Occident, Department of Biology, M. Gaxiola Street and International Road Los Mochis, Sinaloa, Mexico

Karunakaran Maruthachalam,
Department of Plant Pathology, University of California, U.S. Agricultural Research Station, 1636 E. Alisal St., Salinas, CA 93905

James D. McCreight
United States Department of Agriculture, Agricultural Research Service, U.S. Agricultural Research Station, 1636 E. Alisal St., Salinas, CA 93905

Raymundo S. Garcia Estrada
Research Center for Food and Development (CIAD. A. C.), Department of Phytopathology, km 5.5 Culiacan-Eldorado Road, Culiacan, Sinaloa, Mexico

Traditionally, *Erysiphe cichoracearum* (now *Golovinomyces cichoracearum*) and *Podosphaera xanthii*, formerly *Sphaerotheca fuliginea*, have been reported as the causal agents of cucurbit powdery mildew (CPM) (2, 10). These two fungal species have often been misidentified (12), due to similarities in their anamorphic characteristics and the very scarce production of ascomata (13, 16), given that the ascocarpic structure was considered basic for earlier systematics. After considering the fibrosin inclusions in the conidia to differentiate *S. fuliginea* from *E. cichoracearum* in Australia and North America (1, 18), *S. fuliginea* was determined to be the predominant or the only CPM species in the United States (12), although both species can be found infecting the same crops (2, 8, 10).

Molecular analyses demonstrated that some characteristics of the ascomata were not associated with the monophyletic development of the Erysiphaceae, and that *Podosphaera* and *Sphaerotheca* were not separate monophyletic groups (15). Based on the findings of Saenz and Taylor (15), Mori et al. (14) and Takamatsu et al. (17), on the genetic relationships of the Erysiphaceae species, using the Internal Transcribed Spacers (ITS) of the ribosomal DNA (rDNA), Braun and Takamatsu (7) proposed to merge *Podosphaera* and *Sphaerotheca* in the genus *Podosphaera*, and to divide the genus *Podosphaera* in two sections, based on morphological characteristics: *Podosphaera* sect. *Podosphaera* having dichotomously branched ascomata appendages that in *Podosphaera*, sect. *Sphaerotheca* are miceliods.

The Section *Sphaerotheca* was then subdivided in two subsections, in accordance to the size of the peridial cells of the ascomata: small (5 to 25 µm of diameter) in *Sphaerotheca* and large (20 to 55 µm) in *Magnicellulata* (7). The species *P. fusca* and *P. xanthii*, of the Subsection *Magnicellulatae*, were differentiated by Braun et al. (6), based on the size of chasmothecia and the thin walled portion of the asci (oculus), measuring 55 to 90 µm and 8 to 15 µm, respectively, in *P. fusca,* and 75 to 100 µm and 15 to 30 µm, respectively, in *P. xanthii* (3, 6).

Melon (*Cucumis melo* L.), cucumber (*C. sativus* L.), squash (*Cucurbita pepo* L.) and lettuce (*Lactuca sativa* L.) are normally grown in the same greenhouse at the USDA, ARS laboratory, Salinas, Calif. Chasmothecia were for the first time noted in this greenhouse on squash and melons infected by *P. xanthii* during the winter months of 2011 (Fig. 2). Wild lettuce (*L. serriola* L.) PI 491093 plants grown on adjacent benches were infected by *G. cichoracearum* that was not observed to infect the neighboring cucurbits. We characterized morphological and molecular parameters of these two species and cross-inoculated cucumber with *G. cichoracearum*.

Materials and Methods

Plants of several melon, squash and cucumber varieties were planted in a greenhouse on 12 Jan. 2011, in

15 cm x 15 cm x 12 cm deep plastic pots, filled with all-purpose potting soil (Sunland garden, Watsonville, Calif.), watered daily with a solution of 20-20-20 fertilizer (Jack's Classic All Purpose, J.R. Peters, Allentown, Pa.). Air temperature in the greenhouse ranged from 13 to 32 °C (night/day) with a mean of 23 °C. The plants were naturally infected with powdery mildew growing in the same greenhouse. On 22 Mar. 2011, chasmothecia were observed on senescing leaves of 12 of 16 plants of 'Early Summer Golden Crookneck' squash (*C. pepo*) (Fig. 2 A and B), and on one of 10 plants of Iran H and one of eight plants of 'Védrantais', two commonly used *P. xanthii* race differential hosts of melon.

Anamorphic structures were removed from infected leaves using a glossy finish, transparent tape and adhered to a glass microscope slide for microscopic examination. Leaves with chasmothecia were placed under a dissecting microscope where chasmothecia were transferred via needle to drops of 3% KOH on glass slides, covered with a coverslips, and gently pressed to rupture the chasmothecia and liberate the asci. Anamorphic structures, conidia, chasmothecia and asci were observed under a light microscope with an integrated digital camera (Olympus BX60 DP70).

Cross infection of cucumber with G. cichoracearum. Four plants of 'Estrada' cucumber were grown in a CPM-free growth chamber (23 °C; 14/10 h photoperiod; 140 µEm⁻²s⁻²) and inoculated by gently rubbing the leaves with six powdery mildew-infected lettuce leaf discs (1.5 cm diam). Germination and growth of *G. cichoracearum* were observed under a light microscope at 30 h intervals through 120 h post-inoculation, using transparent tape to remove germinated spores.

Molecular characterization of the fungi. Conidia were collected from infected leaves, using a vacuum pump and 200 µL plastic filter tips, and stored in the tips at -20 °C until DNA extraction. The DNA was extracted with the Wizard® Genomic DNA extraction kit (PROMEGA, Madison, Wisc.) and the PCR reactions were made in 20 µL volumes using 1µL of 10 ng·µL⁻¹ DNA template, 1 µL of each primer at a concentration of 10 pmol·µL⁻¹, 10 µL 2X GoTaq® Mastermix (PROMEGA, Madison, Wisc.) and 7 µL nuclease-free distilled water. The primers used were ITS1/ITS4 (18), S1/S2 and G1/G2 (8). The PCR was done using the MJ research PTC-200 (Watertown, Mass.) thermal cycler. The PCR conditions for the ITS1/ITS4 primers were: 94 °C for 5 min, 30 cycles at 94 °C for 1 min, 60 °C for 1 min, 72 °C for 1 min, and a final extension at 72 °C for 10 min. The PCR conditions for the S1/S4 primers were: 94 °C for 5 min, 35 cycles at 94 °C for 1 min, 63 °C for 1 min, 72 °C for 1 min, and a final extension at 72 °C for 10 min. The PCR conditions for the G1/G2 primers were: 94 °C for 5 min, 35 cycles at 94 °C for 1

min, 60 °C for 1 min, 72 °C for 1 min, and a final extension at 72 °C for 10 min. The PCR products were cleaned with the enzyme Exo SAP-IT (Affymetrix/ USB, Cleveland, Oh.) in a thermalcycler for 60 min at 37 °C and 15 min at 80 °C. The cleaned products were diluted ca. 10x to 80 ng·µL⁻¹ and sent for sequencing (McLab, San Francisco, Calif.). The obtained sequences were blasted at National Center for Biotechnology Information (NCBI) for comparison with the *P. xanthii* and *G. cichoracearum* sequences registered at GenBank.

Results and Discussion

The anamorphic structures of powdery mildew (Fig. 1) taken from melon and cucumber leaves matched the descriptions for *P. xanthii*: appresoria indistinct, conidia barrel shaped (dooliform), fibrosin bodies, euoidium type conidiophores, laterally germinated (3-6). Those taken from lettuce leaves matched those of *G. cichoracearum*: euoidium type conidiophores, cylinder shaped conidia, vacuolated with no fibrosin bodies, with length : width > 2, and hyphae with nipple shaped appresoria (3, 4).

The CPM chasmothecia on the melon and squash were ca. 100 µm in diam (Fig. 2 C and D), with big peridial cells, about 12 per plane view, that correspond to the section *Magnicellulatae* (3, 6), contain one ascus (Fig. 2D) with six to eight ascospores (Fig. 2E), and apical openings > 20 mm in diam (Fig. 2E), which indicate that this CPM agent is *P. xanthii* (3, 6).

Conidia of the lettuce powdery mildew pathogen germinated readily on cucumber with a single germ tube from one end, but growth of the germ tubes stopped after 60 h at which time their lengths were < 2x the conidial length (Fig. 3), and a yellowed area was observed in the leaf site of inoculation. These observations were consistent with the generalized characteristics of powdery mildew species germinating on non-host plants (9), and confirmed an earlier negative attempt to infect melon with *G. cichoracearum* obtained from iceberg lettuce grown in an open commercial field (J.D. McCreight, unpublished data).

The PCR amplified products of the pair of primers S1/S2 of two different DNA isolations of *P. xanthii* yielded the same DNA sequence product of 449 bp (GenBank accession JF912574) that had a 99 % identity with sequence AY450960.1 in the same ITS region of *P. xanthii* found on *Fabaceae* in Australia. The same level of identity was observed with ca. 50 powdery mildew accessions, referred to as *P. phaseoli*, *P. balsaminae*, *P. fuliginea* and *P. xanthii* (17).

The sequences obtained from lettuce powdery mildew with the primers ITS1/ITS4 and G1/G2 (GenBank accessions JF951305 and JF951306) had 98 and 100% identity with GenBank sequences AB077688.1 and AB07766.1, respectively, for *G. cichoracearum*. Similar levels of identity were found for *G. orontii*, which infects many families but not members of the Asteraceae of which lettuce is a member (4, 5).

These results confirmed the identity of the CPM pathogen on cucurbits in a Salinas greenhouse as *P. xanthii* based on morphological and molecular characteristics. The powdery mildew pathogen on wild lettuce in the same greenhouse was similarly confirmed as *G. cichoracearum*. Moreover, two *G. cichoracearum* isolates in Salinas (field and greenhouse) may be regarded as representatives of *G. cichoracearum sensu stricto* (4), which is restricted to members of the Asteraceae (5, 11).

USDA is an equal opportunity provider and employer.

Literature Cited

1. Ballantyne, B. 1963. A preliminary note on the identity of cucurbit powdery mildews. Australian J. Sci. 25:360.

2. Bardin, M., Carlier, J., and Nicot, P. C. 1999. Genetic differentiation in the French population of *Erysiphe cichoracearum*, a causal agent of powdery mildew of cucurbits. Plant Pathol. 48 (4):531-540.

3. Bolay, A. 2005. Les oïdiums de Suisse (Erysiphacées). Cryptogamica Helvética 20:1-176.

4. Braun, U. 1987. A monograph of the Erysiphales (Powdery mildews). Hedwigia 89:1-700.

5. Braun, U. 1995. The powdery mildews (Erysiphales) of Europe. Gustav Fischer Verlag, New York.

6. Braun, U., Shishkoff, N., and Takamatsu, S. 2001. Phylogeny of *Podosphaera* sect. *Sphaerotheca* subsect. *Magnicellulatae* (*Sphaerotheca fuliginea* auct. s. lat.) inferred from rDNA ITS sequences-a taxonomic interpretation. Schlechtendalia 7:45-52.

7. Braun, U., and Takamatsu, S. 2000. Phylogeny of *Erysiphe, Microsphaera, Uncinula* (Erysipheae) and Cistotheca, Podosphaera, Sphaerotheca (Cystotheceae) inferred from rDNA ITS sequences–some taxonomic consequences. Schlechtendalia 4:1-33.

8. Chen, R.-S., Chu, C.-C., Cheng, C.-W., Chen, W.-Y., and Tsay, J.-G. 2008. Differentiation of two powdery mildews of sunflower (*Helianthus annuus*) by PCR-mediated method based on ITS sequences. European J. Plant Pathol. 121:1-8.

9. Glazebrook, J. 2005. Contrasting mechanisms of defense against biotrophic and necrotrophic pathogens. Annu. Rev. Phytopathol. 43:205-227.

10. Lebeda, A. 1983. The genera and species spectrum of cucumber powdery mildew in Czechoslovakia. Phytopath. Z. 108:71-77.

11. Lebeda, A., and Mieslerov, B. 2011. Taxonomy, distribution and biology of lettuce powdery mildew (*Golovinomyces cichoracearum sensu stricto*). Plant Pathol. 60:400-415.

12. McCreight, J. D. 2004. Notes on the change of the causal species of cucurbit powdery mildew in the U.S. Cucurbit Genet. Coop. Rpt. 27:8-23.

13. McGrath, M. T., Staniszewska, H., Shishkoff, N., and Casella, G. 1996. Distribution of mating types of *Sphaerotheca fuliginea* in the United States. Plant Dis. 80:1098-1102.

14. Mori, Y., Sato, Y., and Takamatsu, S. 2000. Molecular phylogeny and radiation time of Erysiphales inferred from the nuclear ribosomal DNA sequences. Mycoscience 41:437-447.

15. Saenz, G. S., and Taylor, J. W. 1999. Phylogeny of the Erysiphales (powdery mildews) inferred from internal transcribed spacer ribosomal DNA sequences. Can. J. Bot. 77:150-168.

16. Stone, M. O. 1962. Alternate hosts of cucumber powdery mildew. Ann. Appl. Biol. 50:203-210.

17. Takamatsu, S., Hirata, T., and Sato, Y. 2000. A parasitic transition from trees to herbs occurred at least twice in tribe Cystotheceae (Erysifaseae): evidence from nuclear ribosomal DNA. Mycological Research 104:1304-1311.

18. Thomas, C. E. 1978. A new biological race of powdery mildew of cantaloups. Plant Dis. Rptr. 62:223.

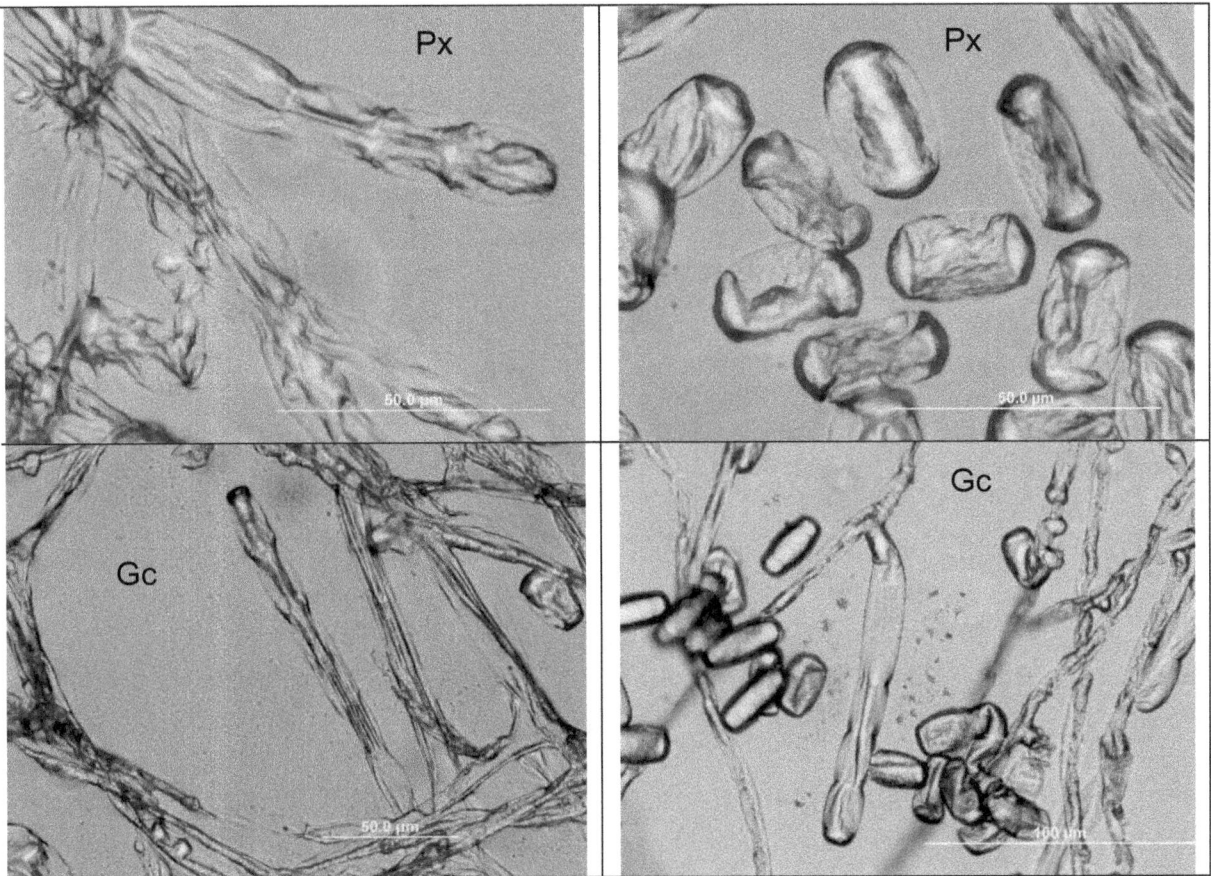

Figure 1. *Podosphaera xanthii (Px) and Golovinomyces cichoracearum (Gc)*, anamorph (L) and conidia (R). *Gc* anamorph from dandelion (*Taraxacum officinale*) and conidia from lettuce (*Lactuca sativa* L.).

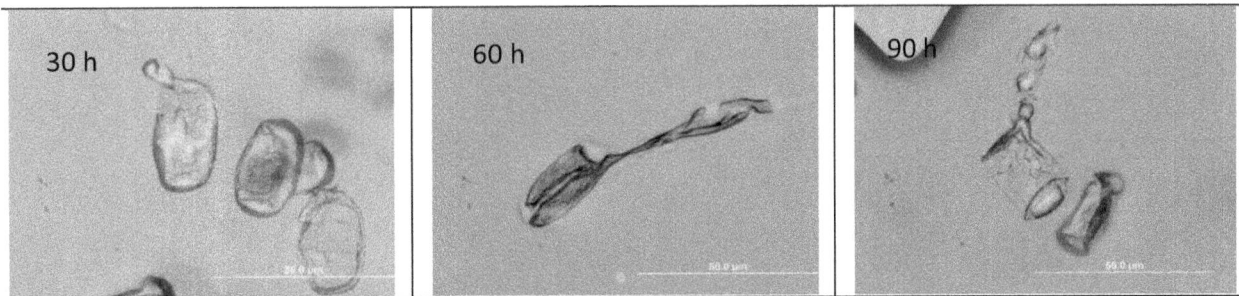

Figure 3. Germination of *Golovinomyces cichoracearum* conidia on 'Estrada' cucumber at 30, 60 and 90 h post-inoculation.

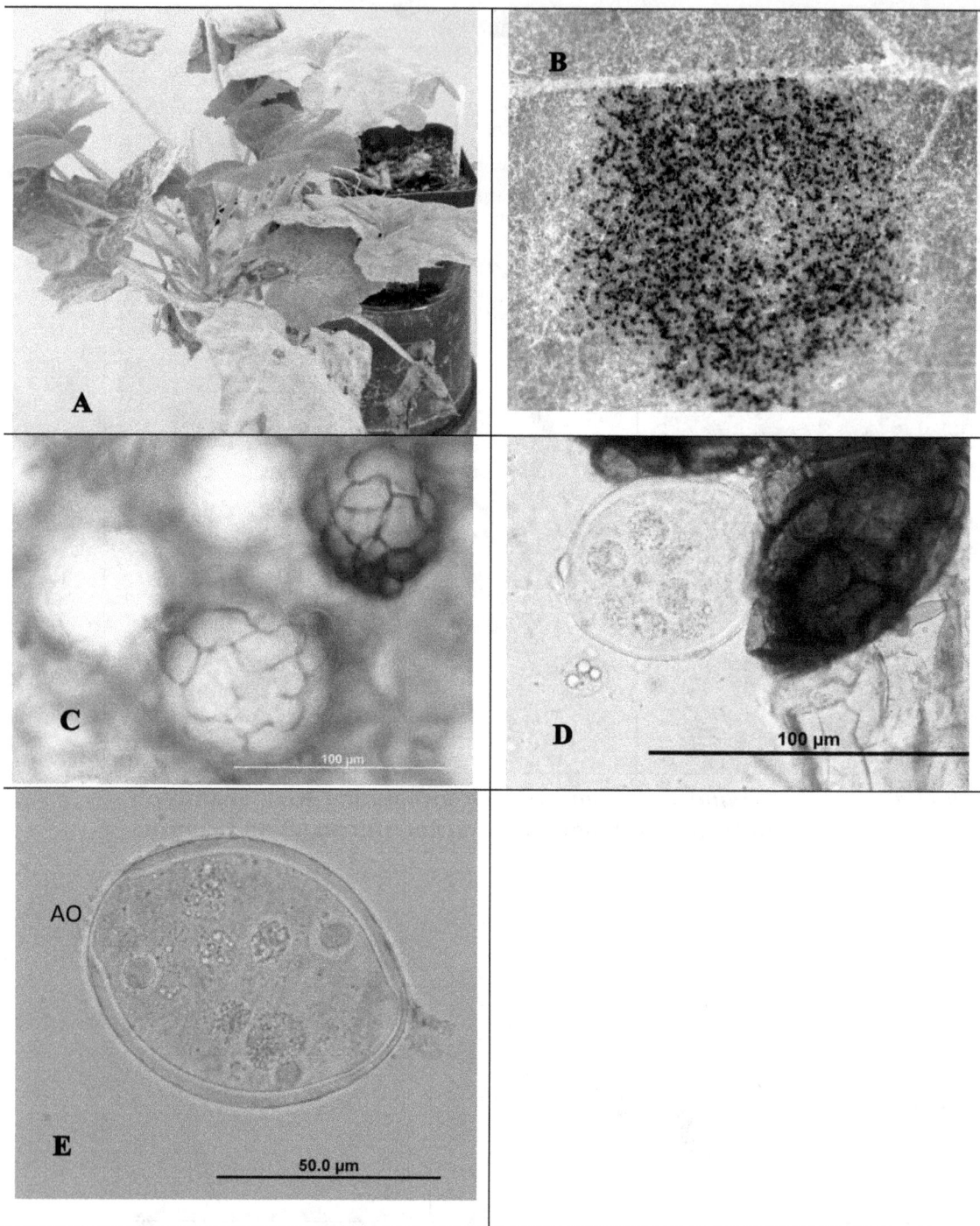

Figure 2. *Podosphaera xanthii* infection (A) and chasmothecia (B) on 'Early Summer Golden Crookneck' squash (*C. pepo*) in a greenhouse, Salinas, Calif. Light microscope views of chasmothecia (C), ruptured chasmothecium with one ascus (D), and (E) ascus with eight ascospores and well developed apical opening (AO).

Melon Trait and Germplasm Resources Survey 2011

James D. McCreight

United States Department of Agriculture, Agricultural Research Service, U.S. Agricultural Research Station, 1636 E. Alisal St., Salinas, CA 93905

The Cucurbit Crop Germplasm Committee (CCGC), which operates under the auspices of the USDA-ARS National Plant Germplasm System (NPGS), is composed of ARS, university and industry scientists, and provides guidance to NPGS on matters relating to cucurbit crop and wild related species. The CCGC is responsible for all cultivated cucurbits (http://www.ars-grin.gov/npgs/cgclist.html).

The CCGC Crop Report is available on-line and is periodically updated. The seed industry and public research communities have undergone major changes since the last update of the melon (*Cucumis melo* L.) section of the CCGC Crop Report. Moreover, changes in market demands and resource and regulatory constraints, further development of molecular technologies, and new production-limiting biotic and abiotic challenges warrant an update of the melon report.

Melon was introduced to North America after centuries of culture and selection throughout Europe, the Middle East, central, eastern and southern Asia, India, and Africa. Recent research supports an Asian origin (1-3) and reports the closest wild species, *C. picrocarpus*, to be in Australia (3).

A two-part survey instrument is, thus, available for input by any person in any country with knowledgeable interest in melon production, utilization, breeding, genetics, or botany (see following). Part 1 is concerned with fruit and plant traits of importance with emphasis on host plant resistance, fungal and viral pathogens in particular. Every region has specific, required combinations of fruit quality traits–they are not included. Part 2 is focused on Germplasm Resources, from "Germplasm," which consists of wild or feral accessions and land races, to varieties and cultivars including F_1 hybrids, and nine types of genetic stocks. Interest in markers and transgenics are also queried.

Download the survey from the website at the bottom of the survey and send via email, or photocopy and send via mail (address above), or fax to the number at the bottom of the survey. All responses will be useful for updating the melon section of the CCGC Crop Report. A summary of the survey will be prepared for inclusion in the next Cucurbit Genetics Cooperative Report.

Literature Cited

1. Renner, S. S., Schaefer, H., and Kocyan, A. 2007. Phylogenetics of *Cucumis* (Cucurbitaceae): *C. sativus* (cucumber) belongs in an Asian/Australian clade far from *C. melo* (melon). BMC Evolut. Biol. 7:58-69.

2. Schaefer, H., Heibl, C., and Renner, S. S. 2009. Gourds afloat: a dated phylogeny reveals an Asian origin of the gourd family (Cucurbitaceae) and numerous oversea dispersal events. Proc. Royal Soc. Bot. 276:843-851.

3. Sebastian, P., Schaefer, H., Telford, I. R. H., and Renner, S. S. 2010. Cucumber (*Cucumis sativus*) and melon (*C. melo*) have numerous wild relatives in Asia and Australia, and the sister species of melon is from Australia. Proc. Natl. Acad. Sci. (U.S.A.) 107(32):14269–14273.

USDA is an equal opportunity provider and employer.

Melon Survey 2011 Part 1: Fruit and Plant Traits of Interest/Needed

Fruit traits
___ Increased shelf life
___ Lightly processed characteristics
___ Increased yield
___ Nutritional value
___ New market types
___ Other: _____
___ Other: _____

Abiotic stress
___ Temperature
___ Salt excess
___ Mineral deficiency
___ Water
___ Other: _____
___ Other: _____

Disease: Fungal
___ Powdery mildew
   ___ *Podosphaera xanthii*
   ___ *Golovinomyces cichoracearum*
___ Downy mildew (*Pseudoperonospora cubensis*)
___ Anthracnose (*Colletotrichum lagenarium*)
___ Fusarium wilt (*Fusarium oxysporum* f.sp. *melonis*)
___ Verticillium wilt (*Verticillium dahliae* and *V. albo-atrum*)
___ Phytopthora crown rot (*Phytopthora capsici*)
___ Pythium (*Pythium* spp.) root and crown rot
___ Alternaria (*Alternaria cucumerina*)
___ Gummy stem blight (*Didymella bryoniae*)
___ Vine decline (*Monosporascus cannonballus*)
___ Other: _____
___ Other: _____

Disease: Bacterial
___ Fruit Blotch (*Acidovorax avenae* subsp. *citrulli*)
___ Bacterial wilt (*Erwinia tracheiphila*)
___ Angular leaf spot (*Pseudomonas syringae* pv. *lachrymans*)
___ Other: _____
___ Other: _____

Disease: Viral

Aphid-transmitted
___ Cucumber mosaic virus (CMV)
___ Cucurbit aphid borne yellows virus (CaBYV)
___ Muskmelon yellow stunt virus (MYSV)
___ Papaya ringspot virus (PRSV) watermelon strain (= WMV 1)
___ Watermelon mosaic virus (WMV) (= WMV 2)
___ Watermelon mosaic Virus-Morocco
___ Zucchini yellow mosaic virus (ZYMV)

Disease: Viral (continued)

Whitefly-transmitted
___ Beet pseudo yellows virus (BPSYV)
___ Cucurbit leaf crumple virus (CuLCrV)
___ Cucurbit yellow stunting disorder virus (CYSDV)
___ Lettuce infectious yellows virus (LIYV)
___ Melon chlorotic leaf curl virus (MCLCV)
___ Melon leaf curl virus (MLCV)
___ Squash leaf curl virus (SLCV)
___ Watermelon curly mottle virus (WCMoV)
___ Other _____
___ Other: _____

Soil-borne/ Seed-borne
___ Muskmelon necrotic spot virus (MNSV)
___ Squash mosaic virus (SqMV)
___ Other _____
___ Other: _____

Other
___ Cucumber green mottle mosaic (CGMMV)
___ Cucumber vein yellowing virus (CVYV)
___ Cucurbit latent virus (CLV)
___ Curly top virus
___ Eggplant mottled dwarf rhabdovirus
___ Kyuri green mottle mosaic virus (KGMMV-YM)
___ Melon rugose mosaic virus (MRMV)
___ Melon vein-banding mosaic virus (MVbMV)
___ Melon yellow spot virus
___ Melon yellowing-associated virus (MYaV)
___ Melon yellows virus
___ Muskmelon yellow spot virus (MYSV)
___ Ourmia melon virus
___ Tobacco ringspot virus (TrSV)
___ Tomato leaf curl virus
___ Zucchini yellow fleck virus (ZYFV)
___ Other _____
___ Other: _____

Insects and Nematodes

___ Sweetpotato whitefly, *Bemisia tabaci* Biotypes: A/B/Q
___ Greenhouse whitefly, *Trialeuroides vaporariorum*
___ Cucumber beetle, *Acalymma trivittatum*, *Diabrotica undecimpunctata undecimpunctata*, and *D. balteata*
___ Leafminer, *Liriomyza sativae* and *Liriomyza trifolii*
___ Green peach aphid, *Myzus persiceae*
___ Melon aphid, *Aphis gossypii*
___ Melon fly, *Myiopardalis pardalina*
___ Root knot nematode, *Meloidogyne* spp.
___ Other: _____
___ Other _____

Download from: http://www.ars-grin.gov/npgs/cgc_reports/melonsurvey

Return to: Jim McCreight, USDA, ARS, 1636 E. Alisal St., Salinas, CA 93905, U.S.A.
Fax 01-831-755-2814; Email: jim.mccreight@ars.usda.gov

Melon Survey 2011 Part 2: Germplasm Resources

Germplasm

___ More germplasm is needed for various characters (as listed on the preceding page).

Germplasm exploration and exchange is:
___ high priority
___ moderate priority
___ low priority
If high or moderate priority, please identify potential countries and areas therein as well as names of in-country contacts who could collaborate in germplasm collection trips or /exchanges.

_____

___ There are adequate germplasm resources available.

Comments on germplasm resources:

_____

Genetic Stocks of interest
___ Recombinant inbred lines (RILs)
___ Near isogenic lines (NILs)
___ Core Collections
___ Test Arrays
___ Mutant Stocks
___ Pocket Collections
___ Haploids
___ Doubled Haploids
___ Tetraploids

Molecular markers are
___ high priority
___ moderate priority
___ low priority

Transgenics *at present* are
___ high priority
___ moderate priority
___ low priority

Transgenics *in the future* are
___ high priority
___ moderate priority
___ low priority

Additional comments

_____

Respondent information (optional, but helpful for any needed clarification)

Name:

Contact information:

Discipline area(s):

___ Breeding   ___ Molecular

___ Pathology   ___ Physiology

___ Entomology   ___ Other _____

Download from: http://www.ars-grin.gov/npgs/cgc_reports/melonsurvey

Return to: Jim McCreight, USDA, ARS, 1636 E. Alisal St., Salinas, CA 93905, U.S.A.
Fax 01-831-755-2814; Email: jim.mccreight@ars.usda.gov

Variability and Correlation among Morphological, Vegetative, Fruit and Yield Parameters of Snake Melon (*Cucumis Melo* Var. *Flexuosus*)

Mohamed T. Yousif
National Institute for Promotion of Horticultural Exports, University of Gezira

Tamadur M. Elamin
Faculty of Agricultural Sciences, University of Gezira

Al Fadil M. Baraka
Agri-technical Group-Khartoum

Ali A. El Jack
Faculty of Agricultural Sciences, University of Gezira

Elamin A. Ahmed
Faculty of Agric. & Nat. Res.-Abu Haraz, University of Gezira

Experiments were conducted at the research farm of the University of Gezira-Wad Medani -Sudan (Latitude 14° 6, N, Longitude 33° 38, E, and Altitude 400m absl), in summer season 2004 and winter season 2004-05. The objectives were to study variability in the local germplasm of snake melon (Cucumis melo var. flexuosus) and correlation among different morphological, vegetative, fruit and yield parameters of snake melon. Five landraces collected in different locations in central Sudan were used in this study. They were subjected to five cycles of inbreeding and selection prior to use in this study. A Complete Randomized Block Design (CRBD), with three replications, was used in the two experiments. Results showed high variability among these inbred lines with respect to earliness (29-45 days), stem length (80.4-152.5 cm), number of secondary branches (7.0-13.5), fruit length (33-90 cm) and leaf pubescence (153-189 hairs/cm²). Results also concluded that yield parameters were much affected by the variation in both genotypes and seasons. Combined analysis of the two seasons indicated positive and significant association for number of primary branches with number of secondary branches (0.43) and stem length (0.35); and number of secondary branches with stem length (0.35), fruit color with sex ratio (0.37); fruit length with groove width (0.43) and female/male ratio (0.42); groove width with stem pubescence (0.43) and sex ratio (0.41). Sex ratio was found to be highly associated with stem pubescence (0.69). In contrast, negative and significant association was found for edible portion with number of primary branches (-0.27) and stem pubescence (-0.38), and groove width with number of straight fruits (-0.28).

Because of the complex taxonomy surrounding the many forms of *C. melo* L. mentioned by Pitrat *et al.* (2000) the *C. melo* var. *flexuous* requires unique morphological description. Some of the descriptors have been added to or modified in the melon descriptor list developed by IPGRI in 1983 such as the Technical Guideline for Melon (TG/104/04); The Cucurbit Genetics Cooperative's Descriptor list for Melon (2003); and GRIN's Descriptors (2000). Other sources taken into account are publications including molecular characterization of melon such as Gomez-Guillamon *et al.* (1985), (1998) and JICA (1995). Although it is of high economic importance, no special descriptors have been developed for snake melon.

Sudan is situated within the region where melon is believed to have originated. Within this region, Sudan could be described as unique with regard to melon ge-

netic resources, both wild and cultivated melon (ElTahir and Mohamed, 2004). Snake melon is among the desirable melon groups in the Sudan. Snake melon is consumed locally as green salad, pickles or stuffed with meat and rice. Farmers used to grow snake melon all over the country throughout the year. The yield of the snake melon crop in Sudan has an average of 20ton/ha (Mirghani and El Tahir, 1997). Variability among local accessions of snake melon was found to be high for morphological and agronomic characters, yield and quality attributes. Recently, a consumer preference of snake melon was developed and they tend to prefer slender fruits with deep grooves and light green outer skin colour, a situation which may threaten the biodiversity among the local cultivated landraces of snake melon in the Sudan. Therefore, priority was given for collection, evaluation and conservation of snake melon and its relatives. The main objective of this study was to estimate variability among collected lines of the cultivated snake melon and correlations among the morphological, agronomic, fruit and yield parameters.

Materials and Methods

Experiments were conducted at the research farm of the University of Gezira-Wad Medani -Sudan (Latitude 14^0 6, N, Longitude 33^0 38, E and Altitude 400m absl), summer season 2004 and winter season 2004-05. The soil is typical of the Central Clay Plain which is characterized by its heavy cracking clay (clay content 58%). Five inbred lines of cultivated landraces of snake melons were used in this study. These were Abu-Haraz which was collected in Gezira state, Silate and Umdum collected in Khartoum State, Farm collected in Kassala State, and Kosti collected in the White Nile State; they were named after the places where they were collected. Pure and stable inbred lines were obtained from the collected landraces after five cycles of inbreeding and selection prior to this study. A Complete Randomized Block Design (CRBD) with three replications was used in these experiments.

The land was disc ploughed, harrowed, and then divided into small growing units (7 x 2.5 m^2). Seeds of the different lines were sown on 15th May and 15th October 2004. Fifty plants of each inbred line were grown in each replication. Three tons of old chicken manure was applied to the soil before the first irrigation and two doses of urea (46% N) were applied 15 and 45 days after sowing at a rate of 40 kg urea/fed. at each application. The crop was irrigated at an interval of 5-7 days during the first month and the irrigation interval was prolonged gradually up to 10 days at the harvesting stage. Hand weeding and chemical control of insect pests and fungal diseases were practiced when-ever it was necessary. Ten plants of each line were randomly taken, tagged and subjected to morphological and agronomic characterization. For estimation of yield the whole number of plants and fruits of each line were considered. Parameters measured in this study included the following:

1- Earliness: The time elapsed from planting to the emergence of the first female flower (days).
2- Sex ratio: Refers to female/male flowers ratio of a plant.

Morphological characteristics such as:
3- Leaf pubescence: Determined by counting number of hairs at the lower surface of the leaf in a disc (1 cm^2) under the microscope.
4- Stem pubescence: Determined by counting number of hairs at the surface of the stem in a disc (1 cm^2) taken at the middle of the stem under the microscope.
5- Leaf colour: Leaves of the different plants were described as dark green, green and light green.

Vegetative characteristics such as:
6- Stem length measured in (cm.) from the soil surface to the tip of the main stem.
7- Number of primary branches: Refers to the number of branches originated from the first node on the main stem.
8- Number of secondary branches: Refers to the number of secondary branches which originated from other nodes rather than the first node of the main stem.

Yield and yield components such as:
9- Fruit length: Determined by measuring the length (cm) of the dorsal side of the fruit.
10- Number of fruits per vine. Determined at the end of the season as total number of fruits harvested from the line divided by the number of plants of the line

Fruit characteristics such as:
11- Fruit colour: It was described using a four categories scale: green, light green, pale green and whitish with the aid of a color chart.
12- Number of straight fruits per vine Shape of fruit was recorded using a three categories scale: straight, semi-curved and curved.
13- Number of ribs: Determined by counting the number of ribs on the outer surface of the fruit.
14- Groove width: measured in cm at the middle of the fruit.
15- Flesh thickness: Measured in cm at the middle of the fruit.

Data were analyzed with MSTAT to calculate ANOVA and correlation coefficients for these traits (Table 1 & 2).

Table 1.

Character	Range of variability	Probability		
		Genotype (A)	Season (S)	Genotype x season (A x S)
Earliness (days)	29-45	0.000	0.28	0.050
Stem length (cm)	80.4-152.5	0.001	0.04	0.006
Number of primary branches	2.8-3.5	0.230	0.35	0.140
Number of secondary branches	7.0-13.5	0.002	0.05	0.213
Number of fruits/vine	5.8-7.2	0.030	0.02	0.050
Fruit length (cm)	33- 90	0.000	0.2	0.034
Groove width (cm)	1.9-2.3	0.000	0.15	0.013
Flesh thickness (cm)	0.6-2.8	0.000	0.01	0.200
Sex ratio (female/male)	0.3-0.5	0.060	0.002	0.010
Leaf pubescence (No of hairs/cm^2)	153-189	0.070	0.02	0.045

Results

Highly significant differences were found among the different lines in the two seasons for earliness (P<0.000), stem length (P= 0.001), number of secondary branches (P= 0.002), number of fruits per vine (P= 0.03), fruit length (P< 0.000), groove width (P< 0.000) and flesh thickness (P< 0.000) as presented in Table 1. The range of variability was high for earliness (29-45 days), stem length (80.44-152.5 cm), number of secondary branches (7.0-13.5), fruit length (33-90 cm), number of fruits per vine (5.8-7.2), groove width (1.9-2.3 cm) and flesh thickness (0.6-2.8 cm). In contrast, no significant differences were found among the different lines for number of primary branches, sex ratio and leaf pubescence. Moreover, differences among seasons were found to be significant for stem length (P=0.04), number of secondary branches (P= 0.04), number of fruits per vine (P= 0.02), flesh thickness (P= 0.01), sex ratio (P= 0.002) and leaf pubescence (P= 0.02). Furthermore, the genotype x season interaction was significant for earliness (P= 0.05), stem length (P= 0.006), number of fruits per vine (P= 0.05), fruit length (P= 0.034), groove width (P= 0.01), sex ratio (P= 0.01) and leaf pubescence (P= 0.045). The color of fruits of the different lines ranged from light green to green.

Stem pubescence of the different inbred lines ranged from few to very few hairs. However, the number of straight fruits was in the range of 20-3, the inbred line 'Farm' recorded the highest number of straight fruits while the inbred line 'Kosti' recorded the lowest number of straight fruits

Association among the different parameters in the summer season, revealed positive and significant correlation of number of primary branches with secondary branches (0.43) and stem length (0.35); number of sec-

ondary branches with stem length (0.35) and total number of fruits per vine (0.31) these results were in a line with those obtained for sweet melon by Taha *et al.* (2003). For morphological characteristics, groove width was found to be associated with sex ratio (0.41) and stem pubescence (0.43); sex ratio with stem pubescence (0.69). Fruit length was found to be highly associated with groove width (0.43) and sex ratio (0.42). Fruit color was found to be highly associated with sex ratio (0.37), the darker the fruit the higher the number of female flowers.

Discussion

Melon (*Cucumis melo* L.) is a polymorphic species especially for fruit characters such as ripening, shape and flesh color. This variability was used by several botanists to subdivide melon into different major groups (Perin *et al.*, 1999). The most recent classification of melon is that given by Pitrat *et al.* (2000). Among these groups snake melon was characterized as monoecious, very long fruit, light green to dark green skin, white flesh, not sweet mature fruit and white seed. Results obtained from this study indicated that phenotypic differences of some characteristics were highly attributed to differences among the genotypes, these included earliness and yield components. Moreover, yield components were also found to be affected by differences due to season. The interaction between the genotype and the season also affected variability in some characteristics such as earliness, sex ratio, stem length, number of fruit per vine, fruit length, groove width and leaf pubescence. These results indicated that both genotype and environmental differences affected snake melon yield and quality and therefore selected lines grown under favorable growing conditions will resulted in high yield and good quality

Some morphological characteristics such as stem pubescence, leaf pubescence were included in this study and they were found to be associated with insect resistance in melon, as mentioned before by IPGRI (2003). Moreover, the ratio of female/male flowers is crucial as it affects pollination especially in monoecious plants such as snake melon. Furthermore, the natural pollination of the different crops in the irrigated sectors in the Sudan is threatened by the appli-

Table 2.

	X$_1$	X$_2$	X$_3$	X$_4$	X$_5$	X$_6$	X$_7$	X$_8$	X$_9$	X10	X$_{11}$	X$_{12}$	X$_{13}$
X$_1$	-	.10	.13	.19	-.11	-.10	-.20	-.27*	-.051	.23	-.38*	.07	-.01
X$_2$		-	-.27*	-.18	-.087	-.021	.37**	-.11	.01	.18	-.29*	.83	-.34*
X$_3$			-	.43**	.03	.084	.42**	.18	.17	.27*	.34*	.07	.07
X$_4$				-	.28*	.11	.41**	.15	.09	.31*	.43**	-.28*	.13
X$_5$					-	-.03	.15	.31*	.04	.29*	.12	-.29*	.04
X$_6$						-	.08	-.03	.08	.04	.17	.072	.09
X$_7$							-	.18	.24	-.16	.69**	-.12	.16
X$_8$								-	.34**	.35**	.15	-.27*	.25
X$_9$									-	.35**	.21	-.05	.31*
X$_{10}$										-	.166	-.03	.14
X$_{11}$											-	-.14	.08
X$_{12}$												-	-.24
X$_{13}$													-

* Correlation is significant at the 0.05 level.
** Correlation is significant at the 0.01 level.
A, Cannot be computed because one of the variables is constant.
X! = Edible portion   X6= Leaf pubescence   X10= Stem length
X2= Fruit color   X7= Sex ratio (female/male)   X11= stem pubescence
X3= Fruit length   X8=Primary branches   X12= Straightnuamber of straight fruits per vine)
X4= Groove width   X9= Secondary branches   X13= Total number of fruits per plant
X5= Fruit color

cation of insecticides. Therefore, breeding hermaphroditic snake melon to enhance pollination was proposed to have good chance for successful and complete pollination which in turn results in a high yield with quality fruits. On the other hand, correlations among fruit length, groove width and straight length can be discussed with the linkage of number of straight fruits per plant with sex type. These associations indicated that hermaphroditic flowers could produce short fruits with broad grooves whereas total number of fruits per vine increase even in the absence of pollinators.

A study of correlation among the vegetative and fruit characteristics in snake melon is almost lacking. The findings of this study revealed association of number primary branches with both stem length and number of secondary branches; fruit length with stem length; groove width with stem length; stem pubescence with both groove width and number of female to male flowers ratio; and fruit thickness with fruit length and fruit color with stem pubescence that can be used in selection breeding of snake melon. Nevertheless, some efforts were exerted to identify this association in sweet melon (*Cucumis melo* var reticulatus), closely related to snake melon by Perin *et al.* (1999) and Taha *et al.* (2003). The latter showed positive and significant association in sweet melon between number of fruits per vine with number of primary branches, number of primary branches with number of secondary branches. These findings were similar to those found for snake melon in this study. Results of correlation among the different growth and yield parameters in snake melon were similar to those found in muskmelon. They also reported strong positive correlation of netting with total soluble solids and flavor. This might explain why snake melon fruits are not sweet and with flat flavor as they are lacking the netting callus on the outer fruit surface. On the other hand, similarity in association among the different characters between sweet and snake melon could indicates also having the same evolution, hybridization and morphological similarity among the different botanical varieties of the species *Cucumis melo* L.

References

El Tahir Ibrahim Mohamed and Mohamed Taha Yousif. 2004. Indigenous melons (*Cucumis melo* L.) in Sudan: a Review of their genetic resources and prospects for use as sources of disease and insect resistance. *Plant Genetic Resources Newsletter* No. 138: 36-42.

Gomez-Guillamon, M.L.; Abadia, J.; Cortes, C. and Nuez, F. 1985. Characterization of melon cultivars. *CGC Rept.* 8:39-40.

GRIN. 2000. Germplasm Resources Information Network. http://www. Ars-grin.gov/npgs/aboutgrin.html.

IPGRI. 2003. Descriptor list for melon (*Cucumis melo* L.). International Plant Genetic Resources Institute, Italy. ISBN 92-9043-597-7. URL: http://www.ipgri.cgiar.org.

JICA. 1995. Cultivation of melon plant genetic resources in: Cultivation and Evaluation of Vegetable Plant Genetic Resources. Technical assistance for genetic resources projects. Ref No. 8 March 1995.

Mirghani, K.A. & El Tahir, I.M., 1997. Indigenous vegetables of Sudan: production, utilization and conservation. In: Guarino, L. (Editor). Traditional African vegetables. Proceedings of the IPGRI international workshop on genetic resources of traditional vegetables in Africa: conservation and use, 29–31 August 1995, ICRAF, Nairobi, Kenya. Promoting the conservation and use of underutilized and neglected crops 16. pp. 117–121.

Perin, C.; Dogimont, C.; Giovinazzo, N.; Besombes, D.; Guitton, L.; Hagen L.and Pitrat M. (1999). Genetic Control and Linkages of Some Fruit Characters in Melon. Cucurbit Genetics Cooperative Report 22:16-18.

Pitrat, M.; Hanelt, P. and Hammer, K.. 2000. Some comments on interspecific classification 0f cultivars of melon. Acta Hort., ISHS (510): 29-36.

Taha, M., Omara, K. and A. ElJack. 2003. Correlation among growth, yield and quality characters in *Cucumis melo* L. *CGC Rept* 26':9-11.

L-Citrulline Levels in Watermelon Cultivars From Three Locations

Angela R. Davis and Wayne W. Fish
Wes Watkins Agricultural Research Laboratory, USDA-ARS, P.O. Box 159, Lane, OK 74555

Amnon Levi
United States Vegetable Laboratory, USDA-ARS, 2700 Savannah highway, Charleston, SC 29414

Stephen King
Department of Horticultural Sciences, Texas A & M University, College Station, TX 77843

Todd Wehner
Department of Horticultural Science, North Carolina State University, Raleigh, NC 27695

Penelope Perkins-Veazie
Plants for Human Health Institute, North Carolina State University, Kannapolis, NC 28083

Additional index words. *Citrullus lanatus*, phytonutrients, amino acids, genetics by environment, quality

Producers of fresh fruits and vegetables face increasing production costs and more intense international market competition. Maximizing marketability by offering high quality produce that is also highly nutritious provides new market niches for crops such as watermelons [*Citrullus lanatus* (Thunb.) Matsum. and Nakai], but germplasm will have to be identified that has enhanced levels of nutrients. Surprisingly, there is little information on how genes can affect the nutritional quality of most fruits and vegetables. This preliminary study was undertaken to determine the importance of genetics versus environment effects in watermelon L-citrulline content, an amino acid that may help regulate blood pressure. Our results suggest that L-citrulline content can vary within a cultivar (one cultivar demonstrated a 0.4 to 4.9 mg/ml fresh sample deviation) even when grown and tested at one location. The data did not indicate a strong varietal difference on the average amount of L-citrulline accumulated (2.4 to 3.4 mg/ml fresh sample); more lines need to be screened to determine if breeding for high L-citrulline germplasm is possible. Location did not appear to significantly increase within-cultivar variation (one cultivar demonstrated a 1 to 4.9 mg/ml fresh sample deviation over two locations), this implies that it may be possible to develop lines with constantly high L-citrulline content across divergent growing environments.

Watermelon is the number two fresh vegetable crop in the world in terms of area harvested and total production (FAOSTAT data, 2007) and has recently been showcased as a healthy food since it is high in the anti-oxidant, lycopene. However, the full nutrient potential of this crop is not known. While watermelon is the leading fruit and vegetable source of lycopene, it also contains a variety of antioxidants and amino acids which likely have additional health-promoting activities. Scientific literature and current medical results indicate that fruits and vegetables contain a host of compounds that appear to work synergistically. Boileau *et al.* (2003) indicated that lycopene alone is not responsible for reducing prostate cancer associated with increased tomato intake. Since tomatoes contain the glycoalkaloid tomatine, a demonstrated anticarcinogen, as well as several well-known phenolic compounds such as quercitin, this finding is not entirely unexpected and demonstrates the importance of maintaining the nutritional value of fruits and vegetables.

Amino acids have well established individual roles in disease prevention. Arginine, an essential amino acid, functions as one of the twenty building blocks of proteins and in free form as a physiologic amino acid. L-citrulline (hereafter referred to as citrulline) is a physiologic amino acid endogenous to most living systems. These amino acids are directly involved in clearing excess metabolic ammonia from the body and are indirectly involved in cardiovascular function, immunostimulation, and protein metabolism (Curtis et al., 2005). Ingested arginine is cleared by hepatic cells, but citrulline is not and can serve as an arginine source in other parts of the body.

Watermelon is rich in citrulline (Tedesco *et al.*, 1984) but varietal differences have not been adequately studied, and effects of growing conditions have not been investigated at all. In an earlier report, fourteen watermelon cultivars were found to contain from 0.42 to 1.83 mg·g^{-1} fresh weight of citrulline, with an average content of 1.5 mg·g^{-1} (Rimando and Perkins-Veazie, 2005). It has been shown that humans can effectively absorb citrulline from watermelon, which also increased plasma arginine levels (Collins *et al.*, 2007; Mandel *et al.* 2005). Recently, subjects consuming watermelon or synthetic citrulline as a drink, combined with exercise, had reduced arterial blood pressure, compared to a placebo (Figureoa *et al.*, 2010). The current study represents a preliminary screen to determine the potential of breeding high citrulline producing watermelons with stable phenotype across multiple environmental locations.

Materials and Methods

Plant material: Five watermelon lines were tested ('Cream of Saskatchewan', 'Red-N-Sweet', 'Tender Sweet Orange Flesh', 'Black Diamond', and 'Dixielee'), three of which (Cream of Saskatchewan, Red-N-Sweet, Tender Sweet Orange Flesh) were grown during the summer of 2008 at two locations in North Carolina, Kinston and Clinton. Black Diamond and Dixielee were grown in 2008 at Lane, OK.

Plants grown in OK were arranged in a randomized complete block design and fertility and care provided as outlined in Motes and Cuperus (1995). All NC samples were nonrandomized and had ten plants per cultivar at each of the two locations. Flesh samples representing each replicate (OK only) and each location were collected from ripe fruit only. Maturity was assessed by external and internal characteristics (i.e., waxyness, tendril death, brix, firmness, seed maturity). A digital refractometer was used to determine Brix. Samples were collected from heart tissue, pureed, and stored at –80°C.

Citrulline quantification: Citrulline was analyzed using a TLC plate methodology which is a slight modification of a Brenner and Niederwieser (1960) method. Citrulline was quantified on the basis of a citrulline standard (Sigma-Aldrich, St. Louis, MO). Briefly, 40 g samples were pureed using a Brinkmann Polytron Homogenizer (Brinkmann Instruments, Inc., Westbury, New York) with a 20 mm O.D. blade. One ml of the liquid puree was centrifuged at 15,800g for 10 min to remove debris. Supernatants were diluted to make a 10% and a 20% solution in deionized water. Ten ul of the diluents were loaded on a 20 x 20 cm silica gel matrix (200 um layer thickness, 5-17 um particle size) TLC plate (Sigma). The spots were air dried, then amino acids were resolved using a solvent (2:1:1 n-butanol: acetic acid: deionized water). Plates were developed with 0.2% ninhydrin in ethanol by baking at 95°C for 5 to 10 min. Densitometric scans of the citrulline spots were visualized and calculated against standards using a Kodak Image station (model 440CF, Eastman Kodak, Rochester, NY). Data in both figures is given as mg/g fresh weight, since 1 ml of watermelon puree is very nearly and consistently 1 g.

Results and Discussion

Though amino acid analysis of protein hydrolysates is generally considered a routine procedure, where as quantification of physiologic citrulline is complicated by presence of glutamine. The method used here estimates citrulline and glutamine together. Since ripe watermelon contains from 3 to 10 times more citrulline than glutamine (Fish and Bruton, 2010), the amount of glutamine in the samples should not adversely affect the reliability of comparisons of citrulline levels among samples.

The five cultivars analyzed had average citrulline content of 2.4 to 3.4 mg/ml fresh sample. These values correlated well with a previous study (overall average 2.4 mg/g) of 14 varieties (Rimando and Perkins-Veazie, 2005). Two cultivars in our study, Cream of Saskatchewan and Tender Sweet Orange Flesh, were also tested in the Rimando and Perkins-Veazie study (2005). The earlier report showed slightly lower citrulline values than we determined (1.0 and 0.5 mg/g, compared to 3.1 and 2.6 mg/g). Ours data showed, on average, higher citrulline values. The difference between the two studies was likely due to the different methods used. This underlies the importance of maintaining one method for comparison between lines. However, we can not rule out that environmental conditions caused differences in the samples between the two studies.

There was no statistical difference between lines in our study. Thus, we were not able to determine if genotype affects average citrulline values. More varieties need to be tested to determine if breeding for high citrulline germplasm is possible.

Our results indicated that within cultivar differences are quite high in some cultivars (one cultivar demonstrated a 0.4 to 4.9 mg/ml fresh sample deviation) even when grown and tested from one location. It might be possible to breed for lines with more stable citrulline expression. Since location did not appear to significantly increase within-cultivar variation (one cultivar demonstrated a 1 to 4.9 mg/ml fresh sample deviation over two locations), it may be possible to breed lines with consistently high citrulline content across widely different growing environments.

Acknowledgements

The authors would like to thank Amy Helms and Cody Sheffield for technical help. This project was partially funded by the National Watermelon Promotion Board and the National Watermelon Association.

Disclaimer

The use of trade names in this publication does not imply endorsement by the USDA of the products named or criticism of similar ones not mentioned.

Presenting author: Angela Davis, USDA-ARS, Wes Watkins Agricultural Research Laboratory, P.O. Box 159, Lane, OK 74555 U.S.A., 580-889-7395, 580-889-5783, angela.davis@lane-ag.org.

Literature Cited

Boileau, T.W, Z. Liao, S. Kim, S. Lemeshow, J.W. Erdman Jr., and S.K. Clinton. 2003. Prostate carcinogenesis in N-methyl-N-nitrosourea (NMU)-testosterone-treated rats fed tomato powder, lycopene, or energy-restricted diets. Journal of the National Cancer Institute. 95:1578-86.

Brenner, M. and A. Niederwieser. 1960. Chromatography of amino acids on thin layers of adsorbents. Experientia. 16:378-383.

Collins, J., G. Wu, P. Perkins-Veazie, K. Spears, P. Claypool, R. Baker, and B. Clevidence. 2007. Watermelon consumption increases plasma arginine concentrations in adults. Nutrition. 23:261-266.

Curtis, E., I. Nicolis, C. Moinard, S. Osowska, N. Zerrouk, S. Benazeth, and L. Cynober. 2005. Almost all about citrulline in mammals. Amino Acids. 29:177-205.

Fish, W.W. and B.D. Bruton. 2010. Quantification of L-citrulline and other physiologic amino acids in watermelon and various cucurbits. Cucurbitaceae 2010. (in press)

Figueroa, A., M.A. Sanchez-Gozalez, P.M. Perkins-Veazie, and B.H. Arjmandi. 2010. Effects of watermelon supplementation on aortic blood pressure and wave reflection in individuals with prehypertension: a pilot study. American Journal of Hypertension. July 8:1-5.

Food and Agriculture Organization of the United Nations. FAOSTAT data. 2007. http://faostat.fao.org/site/567/DesktopDefault.aspx?PageID=567#ancor.

Mandel, H., N. Levy, S. Izkovitch, and S.H. Korman. 2005. Elevated plasma citrulline and arginine due to consumption of Citrullus vulgaris (watermelon). Journal of Inherited Metabolic Disease. 28(4):467-472.

Motes, J. and G. Cuperus. 1995. Cucurbit production and pest management. Circular E-853. Cooperative Extension, Oklahoma State University, Stillwater, OK. 40 p.

Rimando, A.M., and P. Perkins-Veazie. 2005. Determination of citrulline in watermelon rind. Journal of Chromatography A. 1078:196-200.

Tedesco, T.A., S.A. Benford, R.C. Foster, and L.A. Barness. 1984. Free amino acids in Citrullus vulgaris (watermelon). Pediatrics. 73(6):879.

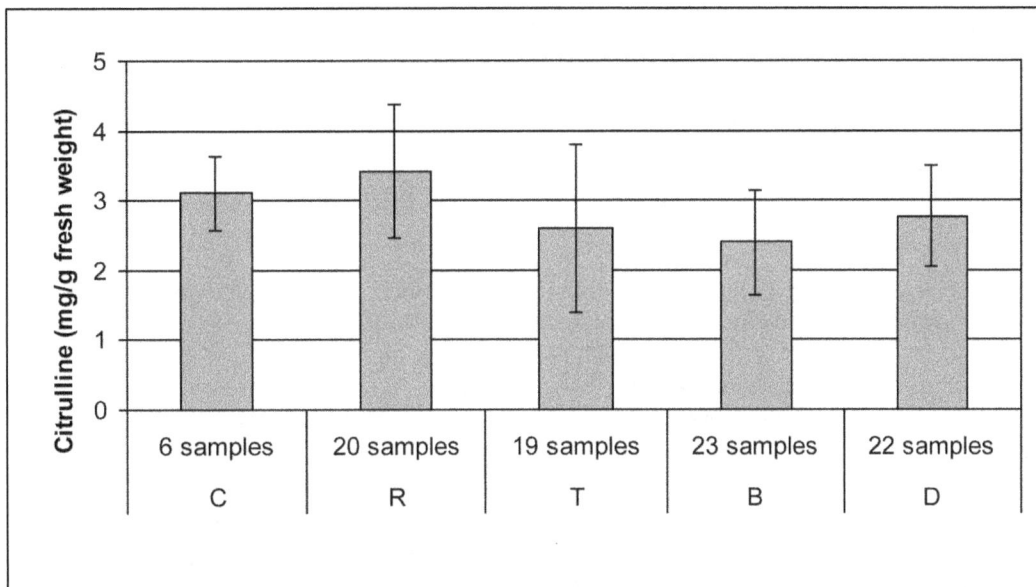

Figure 1. Comparison of citrulline quantities and deviation in five cultivars. The watermelon lines are C, Cream of Saskatchewan; R, Red-N-Sweet; T, Tender Sweet Orange Flesh, B, Black Diamond, D, Dixielee. The number of each cultivar tested is listed. Standard deviation is shown with error bars.

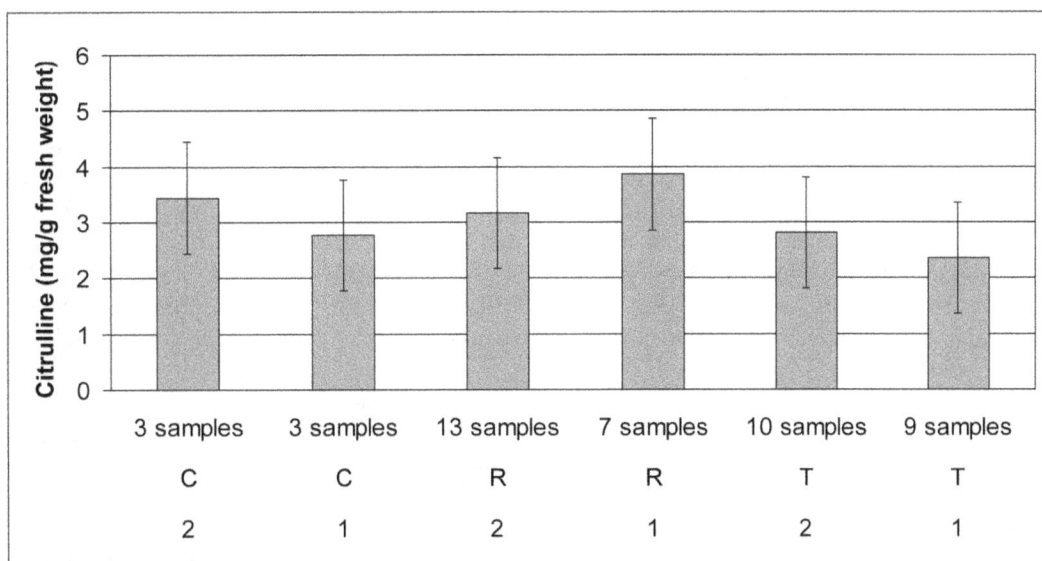

Figure 2. Effect of location on three watermelon lines. The number of each cultivar per location is listed. The watermelon lines are C, Cream of Saskatchewan; R, Red-N-Sweet; T, Tender Sweet Orange Flesh. Location is denoted by a 1, or a 2. Standard deviation is shown with error bars.

Breeding for Yield in Watermelon - A Review

Rakesh Kumar and Todd C. Wehner
Department of Horticultural Science, North Carolina State University, Raleigh, NC 27695-7609

The United States the fourth largest producer of watermelon in the world after China, Turkey, and Iran (5). Yields are highest in China and United States and somewhat lower in the other major producing countries. Watermelon is a major vegetable crop in the U.S. The total area has changed from 76 thousand hectare in 1998 to 65 thousand hectares in 2007 (19). However, production has increased from 1.7 million Mg in 1998 to 1.9 million Mg in 2007 (14). At present, the total value of watermelon production in the United States is $476 million. Over 80% of the watermelon production is concentrated in Arizona, Florida, Georgia, Texas, California, and North Carolina, where the temperatures are warmer and the growing season is longer than states located in northern latitudes (10).

Planned improvement in watermelon was started in the late 19th century in both the public and private sectors (13). In 1954, C.F. Andrus released 'Charleston Gray' with elongate fruit, gray rind, and red flesh. In 1970, C.V. Hall developed 'Allsweet' with similar resistance to 'Charleston Gray', but higher in quality. 'Allsweet' had elongate fruit shape and rind with wide, dark green stripes. J. M. Crall released 'Dixielee', an alternative to 'Allsweet' for its different fruit type and superior quality, and 'Minilee' and 'Mickylee', the first icebox (< 5.5 kg/fruit) cultivars adapted to southeastern U.S. Cultivars that dominated the market in the mid 20th century were open-pollinated ones such as Charleston Gray, Jubilee, Crimson Sweet, and Sugar Baby. By the end of 20th century, hybrids had replaced open-pollinated cultivars for the commercial market. 'Sangria' was the first hybrid developed by T.V. Williams of Rogers NK (now Syngenta) in 1985. The most important change in the watermelon industry is the production of seedless cultivars. O.J. Eigsti released the first seedless watermelon, 'Tri-X-313', in 1962. However, seedless watermelon did not become commercially important until the 1990s due to poor fertility of tetraploid parents used in triploid hybrid seed production. In the U.S., three quarters of the total production is seedless (20), and 'Tri-X-313' is still popular. A recent advance in watermelon breeding was the introduction of mini watermelons that are seedless in the early 21st century. Xingping Zhang developed the first cultivars, sold under the PureHeart™ brand in the U.S. and Solinda™ brand in Europe (13). These watermelons became popular because of their good flavor, crisp texture, and small size.

The yield goal for growers is to harvest at least one load per hectare (51 Mg/ha). Hybrids became popular in 1950s and 1960s. Heterosis is not a big factor in yield for cucurbit crops. Thus, increases in yield cannot be attributed to hybrids over open-pollinated cultivars. Nevertheless, researchers have recorded heterosis in very specific crosses. Some studies reported 10% advantage of the hybrid over the high parent (4, 2). Misra et al. (14) observed that the parents having high per se performance did not always produce hybrids with high hybrid vigor. Gusmini and Wehner (8) reported 22% more fruit yield in F_1 than the best parent. Heterosis up to 47.55 % on per plant fruit yield was reported by in Sel-B x Shipper by Bansal et al. (1). However, they also recorded negative heterosis for fruit yield in some of the crosses. Souza et al. (17) recorded 26 to 41 % mid-parent heterosis for fruit weight in triploid watermelon hybrids. Heterosis might have contributed to yield increase in specific cultivars. In general, the small amount of heterosis for yield in watermelon makes it unnecessary to develop hybrid cultivars, since inbred lines would have similar performance (21). However, hybrids are popular in the seed industry because they provide both protection of the parental lines for intellectual property rights as well as novel traits, such as seedless triploid cultivars. Currently, growers have cultivars of both types available, diploid (seeded) and triploid (seedless) hybrids. Those include the popular diploid hybrids 'Sangria', 'Royal Sweet', 'Fiesta', 'Mardi Gras', and 'Regency' and the popular triploid hybrids 'Tri-X-313', 'Summer Sweet 5244', 'Millionaire', 'Genesis', and 'Tri-X-Shadow' (18).

Numerous yield trials are run each year in the U.S. to evaluate new watermelon hybrids for use as cultivars, but often there are few differences among the entries (7). The question arises as to whether that is due to a lack of genetic diversity for yield in watermelon, or just among the elite, new experimental entries. In the U.S., genetic diversity among watermelon cultivars is narrow because most of them have been derived from just a few original germplasm sources, which includes 'Allsweet'. Gusmini and Wehner (7) tested a diverse set of obsolete inbred cultivars that do not trace to 'Allsweet' type and found that cultivars differed in yield from 36.6 Mg.ha^{-1} in Calsweet to 114.2 Mg.ha^{-1} in Mountain Hoosier. This indicates that genetic variation for yield does exist in the germplasm pool made up of diploid inbred cultivars. Since sources of high yield have been identified, it is important to develop populations using the high yielding cultivars and then use those populations to produce even higher yield.

Currently consumers prefer to have a choice of watermelon fruits from a variety of sizes. Fruit size is a component of yield in cultivated watermelon that is reported as fruit weight, ranging from 1 to 100 kg. Fruit sizes in watermelon are classified as icebox (<5.5 kg), small or pee-wee (5.5-8 kg), medium (8.1-11 kg), large (11.1-14.5 kg), and giant (>14.5 kg) (12). Significant additive, dominant, and epistatic effects have been reported for fruit size, where dominance and dominance-by-dominance effect was largest (16). Brar and Nandpuri (3) found considerable heterosis for fruit size due to partial dominance and overdominance. Gusmini and Wehner (9) recorded low to intermediate estimates of broad- and narrow-sense heritability for fruit size (0.41 and 0.59, respectively).

Yield is a complex quantitative trait, and such traits are typically controlled by many genes, each often having a small effect. In order to improve such traits, it is important to get estimates of heritability, number of genes and gene action. There are several published estimates of broad-sense heritability for yield in watermelon, which are easy to calculate (6, 15). However, in order to develop new inbred lines from segregating populations, it is important to estimate narrow-sense heritability in those populations. Kumar (11) estimated 4-12% estimates of narrow-sense heritability in two watermelon populations indicating complex genetic control of yield. Yield being low heritability trait, selection using single-plant hills would not be effective. A breeding scheme allowing maximum recombination would be useful, and recurrent selection for high yield should be effective.

Literature Cited

1. Bansal, R, B.S. Sooch, and R.K. Dhall. 2002. Heterosis in Watermelon. Environ. & Ecol. 20(4):976-979.

2. Brar, J.S. and A.S. Sidhu. 1977. Heterosis and combining ability of earliness and quality characters in watermelon (*Citrullus lanatus* (Thunb.) Mansf.). J. Res. Punjab Agric. Univ. 14: 272-278.

3. Brar, J.S. and K.S. Nandpuri. 1974. Inheritance of T.S.S., fruit shape, yield and its components in watermelon (*Citrullus lanatus* (Thunb) Mans.). Proc. XIX Intl. Hort. Congr. I. Section VII. Vegetables. Pp. 675-720. [Abstracts].

4. Brar, J.S., and B.S. Sukhija. 1977. Hybrid vigor in inter-varietal crosses in watermelon (*Citrullus lanatus* (Thunb.) Mansf.). Indian J. Hort. 34: 277-283.

5. FAO. 2002. Production Yearbook for 2001. Food Agr. Org. United Nations, Rome.

6. Gill, B.S. and J.C. Kumar. 1986. Variability in genetic stock and heritable components in watermelon. J. Res. Punjab Agric. Univ. 23(4): 583-588.

7. Gusmini, G. and T. C. Wehner. **2005.** Foundations of yield improvement in watermelon. Crop Sci. 45: 141-146.

8. Gusmini, G. and T.C. Wehner. 2004. Heterosis for yield in a watermelon hybrid. Cucurbit Genet. Coop. Rpt. 27:43-44.

9. Gusmini, G. and T.C. Wehner. 2007. Heritability and genetic variance estimates for fruit weight in watermelon. HortScience 42(6): 1332-1336.

10. Hassel, R.L., J.R. Schultheis, W.R. Jester, S.M. Olson, D.N. Maynard, and G.A. Miller. 2007. Yield and quality of triploid miniwatermelon cultivars and experimental hybrids in diverse environments in Southeastern United States. HortTechnology 17(4): 608-617.

11. Kumar, R. 2009. Inheritance of fruit yield and other horticulturally important traits in watermelon [*Citrullus_lanatus* (Thunb.) Matsum. & Nakai. Ph.D Thesis, North Carolina State Univ., Raleigh.

12. Maynard, D.N. (ed.). 2001. Watermelons, Characteristics, production, and marketing. ASHS Press, Alexandria, Virginia.

13. Maynard, D.N., X. Zhang, and J. Janick. 2007. Watermelons: New Choices, New Trends. Chronica Horticulturae 47(4): 26-29.

14. Misra, S.P., H.N. Singh and A. Singh. 1976. Note on heterosis in chilli (Capsicum annum L.). Prog. Hort. 8:61-62.

15. Prasad, L., N.C. Gautam, and S.P. Singh. 1988. Studies on genetic variability and character association in watermelon (*Citrullus lanatus* (Thunb.) Mansf.). Veg. Sci. 15(1):86-94.

16. Sharma, R.R. and B. Choudhary. 1988. Studies on some quantitative characters in watermelon [*Citrullus lanatus* (Thumb.) Matsum. & Nakai] I. Inheritance of earliness and fruit weight. Indian J. Hort. 45: 80-84.

17. Souza, F.F.d., M.A.d. Queiroz, and R.d.C.d.S.Dias. 2005. Heterotic effects in triploid watermelon hybrids. Crop Breeding and Biotechnol. 5:280-286.

18. Tetteh, A.Y. 2008. Breeding for resistance to powdery mildew race 2W in watermelon [Citrullus lanatus (Thumb.) Matsum. & Nakai]. Ph.D. Diss. N.C. State Univ., Raleigh.

19. U.S. Department of Agriculture. 2007. Agricultural statistics for 2006. U.S. Dept. Agr. Washington, D.C.

20. USDA Economic Research Service. 2005. Vegetables and melons situation and outlook yearbook/VGS-2005/July 21, 2005.

21. Wehner, T.C. 2008. Watermelons, p. 381-418. In: J. Prohens and F. Nuez (eds.). Handbook of plant breeding; Vegetables I: Asteraceae, Brassicaceae, Chenopodiaceae, and Cucurbitaceae. Springer Science+Business LLC, New York, NY, 426, p. 17.

Natural Outcrossing in Watermelon - A Review

Rakesh Kumar and Todd C. Wehner

Department of Horticultural Science, North Carolina State University, Raleigh, NC 27695-7609

Watermelon [*Citrullus lanatus* (Thunb.) Matsum. & Nakai var. *lanatus*] belongs to the family *Cucurbitaceae* and subtribe *Benincasinae* (24). Other members of the *Cucurbitaceae* are cucumber, melon, pumpkin, and gourd. The genus *Citrullus* has been divided taxonomically into four species: *C. lanatus* (Syn. *C. vulgaris*), *C. ecirrhosus*, *C. colocynthis*, and *C. rehmii*. Diploid watermelon has 22 chromosomes ($2n=22$, $x=11$) with a genome size of 420 million base pairs (9, 18).It is native to southern Africa, mainly the Kalahari Desert area (2). The secondary center of origin is China. Watermelon can be found growing wild in various parts of western hemisphere, particularly in India (13) and in the Mediterranean region, including Iran and Egypt.

The way plants reproduce depends on their sex expression. This is important in cucurbits because of their different types of sex expression, such as monoecious (staminate and pistillate flowers on the same plant) and andromonoecious (staminate and perfect flowers on same plant) (17). Sex expression in cucurbits besides being genetically controlled is also highly affected by environment (temperature, humidity, light, and nutrition). A single pair of alleles determines sex expression in watermelon. The andromonoecious gene *a* controls monoecious (*AA*) vs. andromonoecious (*aa*) sex expression (9, 14, 15). Watermelon is considered allogamous because both andromonoecious and monoecious sex forms promote cross-pollination. However, both sex forms show varying degrees of self-pollination. The andromonoecious sex form promotes autogamy because of the presence of hermaphroditic flowers, whereas the monoecious sex form promotes allogamy. Allard (1) reported that domesticated cucurbits are more autogamous than allogamous because they originated populations consisting of only a few individuals during domestication. Furthermore, because of their vining growth habit, outcrossing among related individuals may be common, increasing the level of inbreeding, and leading to the purging of deleterious recessive genes. That, in turn, may explain the lack of inbreeding depression in watermelon.

Estimation of natural outcrossing rate is useful for plant breeders especially when experiments are run to estimate components of genetic variance. The genetic structure of plant populations is determined in part by the rate of natural outcrossing. However, consideration of the rate of self-pollination is also important to calculate precise estimates of genetic variances and heritability. In general, individuals within a family in allogamous (cross-pollinated) crops are assumed to be half-sibs, but that is not necessarily the case if self-pollination occurs. As a result of inbreeding, coancestry among half-sibs will be greater than expected (5). Due to self-pollination, variability within families decreases and variability among families increases. The breeding methods applied to self-pollinated crops are distinct from that for cross-pollinated crops. Common methods for crop improvement employed in watermelon are: pedigree breeding and recurrent selection (23). If the natural outcrossing rate is found to be high, watermelon populations can be improved by intercrossing selected families in isolation blocks by recurrent selection. Intercrossing can play an important role in genetic gain.

The factors that influence the rate of natural outcrossing in watermelon are insect pollinators, plant spacing, genotype (cultivar), and climatic conditions. Cross-pollination in watermelon is mediated by honeybees (*Apis mellifera* L.) and bumblebees (*Bombus impatiens* Cresson) that visit the flower to collect pollen and nectar (4, 7, 12). Although >85% of watermelon pollinators are honey bees, bumble bees have been reported to be a better pollinator than honey bees in watermelon (21). Most of the pollen is removed in 2 hours after anthesis in watermelon (20) by pollinators. Gingras et al. (8) suggested that a single visit is enough to induce fruiting. The movement of insect pollinators in a field is strongly directional, with pollinators moving to the nearest neighboring flowers within the same row (3, 10, 22, 26). In addition to insect pollinators, the outcrossing rate is also reported to be influenced by staminate flower and pollen production as affected by the genotype and environment. Pollen movement was restricted to 3 m from the donor plant in muskmelon (10) and 2 to 3 m in cucumber (11). Stanghellini and Schultheis (19) reported variability in pollen grain production in 27 watermelon cultivars.

In watermelon, the rate of natural outcrossing (measured between-row only) was near zero for rows separated by 6 m or more (16) and averaged 0.8% for rows 3 to 6 m apart. Walters and Schultheis (23) recorded an outcrossing rate near to zero in plants spaced more than 10 m apart. Ferreira et al. (5) reported an outcrossing rate of 65% and inbreeding coefficient as high as 0.41 in andromonoecious families of watermelon. When averaged over monoecious and andromonoecious families, the outcrossing rate was 77% (5, 6). However, these authors did not report the plant spacing adopted in the experiment. The rate of natural outcrossing has been measured for cucumber families planted in isolation blocks. Wehner and Jenkins (24) reported that natural outcrossing rate (mean and range over replications) was 36% (29-43%) cross-, 17% (0-42%) sib-, and 47% (23-77%) self-pollination. Thus, 64% of pollinations were self- or sib-, but not crosspollination among families. Watermelon was expected to have a mixed mating system since it is similar to cucumber in plant growth and sex expression. Moreover, there was no significant inbreeding depression in watermelon (23), indicating a high rate of self-pollination in the species. Self-pollination can occur in both monoecious and andromonoecious populations (6). Allard (1) suggested that cucurbits evolved as small populations in nature, thus having high levels of inbreeding.

Watermelon breeders often calculate estimates of genetic variance and covariance among family members (e.g. half-sibs) in their populations. Estimates are often miscalculated if the mating system is not well studied. The coancestry of individuals is higher when parents are spaced widely, due to an increase in self-pollination. Genetic variance may be calculated as: $\sigma^2_G = (1+F) \sigma^2_A + (1-F) \sigma^2_D + 4FD_1 + 4FD_2 + F(1-F)H$, where σ^2_G is genetic variance (25). F, σ^2_A, σ^2_D, D_1, D_2, and H are inbreeding coefficient, additive variance, dominance variance, covariance between additive and homozygous variance effects, variance of homozygous dominance effects, and measure of inbreeding depression, respectively.

Further studies are needed to measure the amount of self- and cross-pollination in watermelons at different plant spacings. If the rate of self-pollination is high in widely-spaced plants, it would be possible to use open pollination for the initial generations of selection during the development of lines without the expense of controlled pollination.

Literature Cited

1. Allard, R.W. 1960. Principles of plant breeding. Wiley, N.Y.

2. Bailey, L.H. 1949. Manual of cultivated plants. McMillan Company, New York, p 950.

3. Cresswell, J.E., A.P. Bassom, S.A. Bell, S.J. Collins, and T.B. Kelly. 1995. Predicted pollen dispersal by honey-bees and three species of bumble-bees foraging on oil seed rape: A comparison of three models. Funct. Ecol. 9: 829-841.

4. Delaplane, K.S. and D.F. Mayer. 2000. Cross pollination by bees. CAB Intl., Wallingford, UK.

5. Ferreira, M.A.J. da. F., M.A. de Queiroz R. Vencosky, L.T. Braz, M.L.C. Viera and R.M.E Borges. 2002. Sexual expression and mating system in watermelon: implications for breeding programs. Crop Breeding & Appl. Biotechnol. 2(1): 39-48.

6. Ferreira, M.A.J. da. F., R. Vencosky, M.LC. Viera, and M.A. de Queiroz. 2000. Outcrossing rate and implications for the improvement of a segregating population of watermelon. Acta Horticulturae (510): 47-54.

7. Free, J.B. 1993. Insect pollination of crops. 2nd Ed. Academic Press, London, UK.

8. Gingras, D., J. Gingras, and D. Oliveira. 1999. Visits of honeybees (Hymenoptera: Apidae) and their effects on cucumber yields in the field. J. Econ. Entomol. 92(2): 435-438.

9. Guner, N. and T.C. Wehner. 2004. The Genes of watermelon. HortScience 39(6): 1175-1182.

10. Handel, S.N. 1982. Dynamics of gene flow in an experimental population of *Cucumis melo* (Cucurbitaceae). Amer. J. Bot. 69: 1538-1546.

11. Handel, S.N. 1983. Contrasting gene flow patterns and genetic subdivision in adjacent populations of *Cucumis sativus* (Cucurbitaceae). Evolution 37: 760-771.

12. McGregor, S.E. 1976. Insect pollination of cultivated crops. U.S. Dept. Agr. Res. Serv. Agr. Hdbk. 496.

13. Peter, K.V. 1998. Genetics and Breeding of Vegetables. Indian Council of Agr. Res. p129.

14. Rhodes, B. and F. Dane. 1999. Gene list for watermelon. Cucurbit Genet. Coop. Rpt. 22: 61-77.

15. Rhodes, B. and X. Zhang. 1995. Gene list for watermelon. Cucurbit Genet. Coop. Rpt. 18: 69-84.

16. Rhodes, B.B., W.C. Adamson, and W.C. Bridges. 1987. Outcrossing in watermelons. Cucurbit Genet. Coop. Rpt. 10: 66-68.

17. Robinson, R.W., H.M. Munger, T.W. Whitaker, and G.W. Bohn. 1976. Genes of cucurbitaceae. HortScience 11: 554-568.

18. Shimotsuma, M. 1963. Cytogenetical studies in the genus *Citrullus*. VII. Inheritance of several characters in watermelon. Jpn. J. Breeding 13: 235-240.

19. Stanghellini, M.S. and J.R. Schultheis. 2005. Genotypic variability in staminate flower and pollen grain production of diploid watermelons. HortScience 40(3): 752-755.

20. Stanghellini, M.S., J.R. Schultheis, and J.T. Ambrose. 2002. Pollen mobilization in selected Cucurbitaceae and the putative effects of pollinator abundance on pollen depletion rates. J. Amer. Soc. Hort. Sci. 127(5): 729-736.

21. Stanghellini, M.S., J.T. Ambrose, and J.R. Schultheis. 1998. Seed production in watermelon: A comparison between two commercially available pollinators. HortScience 33(1): 28-30.

22. Walters, S.A. and J.R. Schultheis. 2009. Directionality of pollinator movements in watermelon plantings. HortScience 44(1): 49-52.

23. Wehner, T.C. 2008. Watermelons, p. 381-418. In: J. Prohens and F. Nuez (eds.). Handbook of plant breeding; Vegetables I: *Asteraceae, Brassicaceae, Chenopodiaceae*, and *Cucurbitaceae*. Springer Science Business LLC, New York, NY, 426, p. 17.

24. Wehner, T.C. and S.F. Jenkins, Jr. 1985. Rate of natural outcrossing in monoecious cucumbers. HortScience 20: 211-213.

25. Weir, B.S and C.C. Cockerham. 1976. Two-locus theory in quantitative genetics. P. 247-269. In: E. Pollak, O. Kempthorne, and T.B. Bailey, Jr. (ed.). Proc. Intl. Conf. on Quantitative Genet. Ames, Iowa. 16-21 Aug. 1976. Iowa State University Press, Ames.

26. Zimmerman, M. 1979. Optimal foraging: A case for random movement. Oecologia 43: 261-267.

Characterization of M1 Generation Of Polyploids in Watermelon Variety 'Sugar Baby'

T. Pradeepkumar

Associate Professor, Department Of Olericulture, College Of Horticulture, Vellanikkara, P.O. Kau, TRICHUR,KERALA, INDIA 680656 pradeepkau@gmail.com

Watermelon [*Citrullus lanatus* (Thunb.) Matsum and Nakai] exhibit great variation in seed morphology, particularly in seed size and shape, and in seed coat tissue texture, and color. However, in recent years, seedless watermelon are more acceptable to consumers, and seed companies make great efforts in breeding for high quality seedless watermelon varieties. Kihara (1951) produced triploid seedless hybrid watermelon by crossing tetraploid (4n) and diploid (2n) plants. A large number of triploid varieties have been developed in Asia and North America. However, in contrast with the USA, seedless watermelons have not been popular in India. There is a need to develop tetraploid breeding lines suitable for consumer needs in India. Breeding stable tetraploid line with adequate fertility is a major challenge in triploid watermelon production (Mohr, 1986).

Compton *et al.* (1996) developed an easy technique for the early identification of putative tetraploid watermelon plants. In tetraploid plants, the guard cells of the stomata contain a high number of chloroplasts compared with these in diploid plants. Here, a procedure has been optimized to induce polyploidy in the commercial watermelon cultivar Sugarbaby, and the resultant M1 generation plants were morphologically characterized.

Materials and Methods

Polyploidy was induced in the seeds and seedlings of the watermelon variety Sugarbaby using colchicine solution (0.1 % and 0.5%). The seeds of Sugarbaby were soaked in clean water for 6 hrs and then soaked for 24 hrs in a solution containing 0.1, 0.5, 1, 1.5, or 2.0% colchicine. The seeds were briefly rinsed before sowing in polybags filled with sand. For seedling treatment seeds were sown in polyhouse. Once seedlings emerged, a drop of colchicine (0.1 or 0.5%) was applied to the shoot apex between the cotyledons. The chemical solution was applied to the growing point during morning hours for three consecutive days. The plants derived from seed and seedling treatments were examined with respect size of stomata guard cells and number of chloroplasts in each guard cell.

Results and Discussion

Seeds failed to germinate when exposed for 24 hours to a concentration of 1, 1.5 and 2 % colchicine. Variable response to high concentration of colchicine was reported by Jaskani (2005). Giant guard cell was observed in plants generated through 0.1% colchicine seed treatment and 0.5% colchicine seedling treatment (Table 1). Polyploid plants had a higher chloroplast number, ranging from 17-22, while the diploid (control) plants had about 12 chloroplasts in the stomata guard cells. Chloroplast counting is an efficient indicator of polyploidy during early phase of plant growth (Jaskani *et al.*, 2005) and a higher chloroplast number is an indicator for higher ploidy. In general, polyploids took more days to form male and female flowers (Table 2). Slower growth rate and delayed appearance of shoots and flowers were observed in colchicine treated seedlings. The same trend was also reflected in the days taken to harvest.

Plants generated following treatment of seedlings with 0.5 % colchicine produced large pollen grains (40.35 µ). Consequently, the polyploidy plants produced smaller fruits with a small number of seeds (Table 3). The number of seeds produced in plants treated with 0.5 % colchicine (15-46 seeds) is significantly lower than that produced in the diploid control plants (430-530 seeds). In general, plants regenerated from seeds treated with 0.1 % colchicine or from seedlings treated with 0.5 % colchicine exhibited polyploidy characters. On the other hand, seedling treated with 0.1 % colchicine produced chimeral growth and normal fruits with high seed numbers. Variability in colchicine tolerance by watermelon lines was observed in a previous study (Jaskani *et al.*, 2004). The variable response of genotypes to colchicine indicates that optimal colchicine concentrations may vary in treating watermelon genotypes.

Although tetraploid watermelon plants are morphologically distinct, they may not be stable and produce mixoploid vines on the same plant. Rose *et al.* (2000) reported that mixoploid plants in which more than 40% of the cells were tetraploid could be mistaken for full tetraploids but that mixoploids with 10–30% tetraploid cells usually resembled diploid plants. However, chloroplast counts and other morphological features could be helpful in successive generations to develop pure line. The results of this study underscore the need for continual observation for distinguishing polyploids from diploid plants and presence of mixoploids calls for a cautious approach in breeding polyploid lines. Overall, ploidy level can be determined by examining epidermal tissue from the lower leaf surface, without the requirements of specific equipment and high expenditure.

Literature Cited

Compton, M.E, D.J. Gray, and G.M. Elmstrom. 1996. Identification of tetraploid regenerants from cotyledons of diploid watermelon cultures *in vitro*. Euphytica 87:165-172.

Jaskani, M.J., W.K. Sung, and H.K. Dae. 2005. Flow cytometry of DNA contents of colchicines treated watermelon as a ploidy screening method at m1 stage. Pakistan Journal of Botany 37: 685-696.

Jaskani, M.J., S.W. Kwon, G.C. Koh, Y.B. Huh and B.R. Ko. 2004. Induction and characterization of tetraploid watermelon. Journal of Korean Society of Horticultural Science 45: 60-65.

Kihara, H. 1951. Triploid Watermelons Proceedings of American Society of Horticulture Science 58 : 217-230.

Mohr, H.C. 1986. Breeding Vegetable Crops AVI Publishing Company, Westport, 584pp

Rose, J.B., J. Kubba and K.R. Tobutt. 2000. Chromosome doubling in sterile *Syringa vulgaris* x *S.pinnatifolia* hybrids by *in vitro* culture of nodal explants. Plant Cell Tissue and Organ Culture 63: 127-132.

Sl no	Treatment		Length (Micron)	Width (Micron)	No.of chloroplast/cell
			Leaf Guard cell		
1	0.1% seed	1	29.55±1.2	21.21±1.7	19±2.0
		2	28.82±1.5	22.99±1.9	19±1.2
		3	26.99±1.3	22.79±1.5	18±1.1
		4	29.49±1.7	22.58±1.3	19±1.2
2	0.1% Seedling	1	29.85±1.3	21.17±1.3	18±0-5
		2	30.55±1.2	22.70±1.1	17±0.3
		3	27.06±1.2	20.47±1.2	17±1.0
		4	25.19±1.1	21.50±1.2	17±1.2
3	0.5% Seedling	1	28.76±1.2	21.48±1.3	22±0.7
		2	26.59±1.2	21.67±1.2	22±1.0
		3	29.05±1.2	21.61±1.3	22±0.7
		4	29.44±1.2	23.96v1.3	22±0.6
4	Control	1	20.52±1.4	15.86±1.0	12±0.3

Table 1 : Leaf characters of plants produced from seeds or seedlings treated with colchicine versus untreated control plants. (values are mean±SD , n=12)

Sl no	Treatment		Size of pollen grain (Micron)	Days taken to open first male flower	Days taken to open first female flower	Days taken to harvest
1	0.1 seed	1	35.0±0.2	57	69	137
		2	34.54±0.3	57	69	137
		3	35.09±0.4	52	69	131
		4	33.07±0.2	52	69	131
2	0.1 seedling	1	32.69±0.2	56	60	145
		2	31.26±0.1	50	62	153
		3	32.20±0.1	58	63	140
		4	32.00.±0.2	63	65	142
3	0.5 seedling	1	37.46±1.0	57	63	116
		2	33.05±1.1	60	68	153
		3	38.04±1.0	60	68	153
		4	40.35±1.1	61	61	166
4	Control	1	31.74±1.0	29	32	82

Table 2 : Biometric characters of plants produced from seeds or seedlings treated with colchicine versus untreated control plants. (values are mean±SD , n=12)

Sl no	Treatment		Fruit					
			Rind thickness (cm)	Wt (Kg)	Perimeter (cm)	TSS	Seed no	Avg.wt of 10 seeds
1	0.1% seed	1	2	1.51	45	7	32	0.2
		2	1.5	1.21	34	7	51	0.2
		3	1.5	2.64	54	8.3	506	0.3
		4	1.5	1.50	40	8	15	0.2
2	0.1% seedling	1	1.5	4.73	65	9.3	596	0.3
		2	1.5	1.50	45	7	323	0.3
		3	1.5	1.81	49	7.8	521	0.3
		4	1.5	2.51	55	7.8	340	0.3
3	0.5 % Seedling	1	1	0.50	30	7.5	34	0.25
		2	1.4	0.96	40	8	18	0.3
		3	1	0.52	33	8	15	0.3
		4	1	1.10	38	10.3	46	0.4
4	Control	1	1.5	3.20	55	7	530	0.3

Table 3: Fruit characters of plants produced from seeds or seedlings treated with colchicine versus untreated control plants.

A "Hull-less" Seed Trait of *Cucurbita maxima* Duch. in Accession BGH 7653

José Raulindo Gardingo
Departamento de Fitotecnia e Fitossanidade, Universidade Estadual de Ponta Grossa, Av. Carlos Cavalcanti, 4748, CEP 84030-900, Ponta Grossa, PR, Brasil. e-mail: jrgardin@uepg.br

Derly José Henriques da Silva, Vicente Wagner Dias Casali and Izaias da Silva Lima Neto
Departamento de Fitotecnia, Universidade Federal de Viçosa, Campus, CEP 36570-000, Viçosa, MG, Brasil.

Roseli Aparecida Ferrari
CCQA / ITAL, Av. Brasil, 2880, CP 139, CEP 13070-178, Campinas, SP, Brasil.

The consumption of seeds of *Cucurbita* sp. as food is common in many cultures throughout the world. The hull-less seeds are served as snackseed and their oil can be used in salads. An assessment of pumpkin (*Cucurbita maxima* Duch.) from the Germplasm Bank of Vegetables (BGH) of the *Universidade Federal de Viçosa* (Federal University of Viçosa) revealed the production of hull-less seeds by a mutation in accession BGH 1518 (designated BGH 7653). Thus, this work aimed at the physical-chemical characterization of the hull-less seeds identified, by comparing them with seeds with normal integument from the original accession. The weight of one thousand seeds was similar among the genotypes; however, the dry seed yield per fruit of the original sample was higher. The chemical composition of the normal seeds was 41.75% of oil, 28.64% of proteins and 3.74% of ash, whereas the corresponding composition of the mutant BGH 7653 was respectively 46.70%, 37.93%, and 4.5%. The accession BGH 7653 has potential for breeding programs concerning the consumption of roasted seeds, the oil production for salads, and the use for medicinal and industrial purposes.

Keywords: oil, protein, ash, pumpkin, naked seed.

Introduction

The seeds of *Cucurbita* sp. have been used as food and medicines (1,3,14), because they are excellent sources of proteins (32-44%), oil (34-50%) and vitamin E (1,2,3,5,6,7,8,10,12, 14). In Austria, the extraction of oil from the seeds of *Cucurbita pepo* L. has been carried for more than 100 years. Extraction from hulled seeds is difficult, but the extraction became easier and more efficient with the use of a natural mutant, which hindered the lignification of the hull (3,6,7,8,11,12). Currently, the oil production for salad using hull-less seeds or naked seeds of *C. pepo* are found mainly in Austria, Hungary, Slovenia and Yogoslavia (3,4,8,12,13). The seeds of *C. maxima* contains over 70% of unsaturated fatty acids comprised mainly by linoleic and oleic acids (1,10). The commercial exploitation of genotypes that produce hull-less seed has not been reported for Brazil. However, broad genetic variability is preserved in the Germplasm Bank of Vegetables (BGH) from the Federal University of Viçosa (UFV), which preserves more than 295 accessions of pumpkins (*C. maxima* Duch.) (9). Thus, this paper aims to describe hull-less seeds of *Cucurbita maxima* from the accession BGH 7653, which has potential for integrating the breeding program for human nourishment as well as for oil production. To our knowledge, this is the first report of the hull-less trait in a Cucurbit species other than *C. pepo*.

Materials and Methods

The experiment was carried out at *Horta Velha* located at 600 m of altitude (latitude 42° 52' 53'' S, longitude 20° 45' 14'' W), Viçosa, Minas Gerais state, Brazil from February to August 2009. An assessment of the seeds of BGH 1518 accession from the Germplasm Bank of Vegetables from the Federal University of Viçosa revealed a hull-less type (designated BGH 7653) apparently arising by spontaneous mutation. The sowing occurred on February 18th 2009 and the transplantation was carried on March 09th 2009. During the culture, normal production and phytosanitary treatments were applied. Plants were sprinkler irrigated. Controlled pollinations were carried out between plants from the same accession, with their flowers protected immediately after pollination using paper bags. The fruits were harvested

45 and 55 days after pollination during the winter, and stored for 45 days. They were weighted before the manual removal of the seeds, and immediately gently washed on a steel sieve with running water to remove the mucilage. After drying, they were solar irradiated for one day, and plump seed were separated from shrivelled seeds. The seeds with moistures around 10% humidity were packed into plastic bags and stored at 5°C. One sample was used to measure the physical characteristics and chemical composition of the seeds. Regarding the physical characteristics of the seeds, ten seeds were randomly chosen from each fruit for length, width and thickness measurements using a digital calliper. Additionally, some seeds were randomly measured to determine hull thickness. We used twelve fruits harvested from different plants from the original accession with hulled seeds of BGH 1518 and only two fruits from the same plant of the hull-less mutant (BGH 7653).

The chemical composition determination of the seeds were carried out in duplicate through the following methods: (a) the dry matter, by drying the seeds at 105 °C for 24 hours; (b) the protein content, by the method of micro Kjeldahl using the factor 6.25; (c) lipid content, by the Soxhlet device using hexane as solvent; and (d) the ash incinerating at 600 °C in a muffle furnace for 4 hours.

The software Microsoft Excel was used for the data analysis and the results expressed the mean and standard deviation (SD) from different accessions.

Results and Discussion

Variation was observed for the seed yield, physical characteristics and for chemical composition of hull-less seeds when compared to the seeds with normal integument (Table 1) therefore demonstrating the potential for the breeding program to enhance the value of the product and to create new niches in the market, which would create more jobs and would provide more income to farmer.

The weight of the fresh fruit was 2.819 kg (SD ± 1.54) for the accession BGH 1518 and 3.139 kg (SD ± 0.04) for the mutant BGH 7653.

The number of seeds obtained per fruit, the dry weight of seeds per fruit, and the weight of one thousand seeds were 498.38 (SD ± 248.76), 78.94g (SD ± 36.42) and 164.91g (SD ± 37.30) for the accession BGH 1518 and 268.04 (SD ± 51.36), 42.707g (SD ± 10.90) and 158.34g (SD ± 51.36) for the mutant BGH 7653, respectively. The difference between the total dry weight of the accession and the mutant is mainly due to the number of seeds per fruit. The values regarding the number of seeds per fruit and the total dry weight of the mutant BGH 7653 were

higher than those obtained by (7) for the four selected lines of *C. pepo* "hull-less", showing therefore the potential of hull-less seeds as snackseeds. The weight of one thousand seeds of the genotypes assessed was lower than the 12 cultivars of *C. maxima* used by Stevenson *et al.* (10).

The ratio between the dry weight of the seeds and the fresh weight of the fruits was 2.95% (SD ± 0.94) for the accession BGH 1518 and 1.36% (SD ± 0.36) for the hull-less mutant BGH 7653. Such difference mainly reflects the number of seeds obtained per fruit, where the mutant BGH 7653 produced almost half of the number of the original genotype. A 20-year program focused on selection in *C. pepo* to increase hull-less dry seeds mass compared to fresh fruit weight achieved an increase from 1.5% to approximately 3.0% (13).

The length, width and thickness of the seeds were 16.8mm (SD ± 1.39), 8.96mm (SD ± 0.96) and 3.13mm (SD ± 0.49) for the accession BGH 1518 and 15.5mm (SD ± 0.42), 9.16 mm (SD ± 0.62) and 2.81 (SD ± 0.16) for the mutant BGH 7653, respectively. The genotypes have similar measurements.

The thickness of the seed depends on its hull and embryo thickness. According to Andres (3), the testa of seed varies in thickness and hardness, and thinner testa is tolerated when ground or roasted. Stuart and Loy (11) measured several layers of the testa and found that the dry hull-less seeds have minimal dimensions. Thus, the measures from the normal type showed 0.43 mm in dimension while the hull-less mutant variety 'Tricky Jack' was 0.20 mm. We observed variable thickness in different plants of BGH 1518, whose white hulled seeds showed a mean thickness of 0.26 mm (SD±0.03) and the brown hulled seeds 0.63 mm (SD± 0.18).

The accession BGH 7653 had similarities with the hull-less mutant described by Stuart & Loy (11), Winkler (12) and Fruhwirth & Hermetter (6). The mutant showed the embryo cover comprised of a rudimentary testa (Figure 1).

The moisture level of the seeds was 8.77% (SD ± 1.03) for the accession BGH 1518 and 8.97% (SD ± 2.32) for the mutant BGH 7653. Such values were similar to those observed for *C. maxima* by Cerqueira *et al.* (5), but considerably higher than those obtained by Alfawaz (1) and Amoo *et al.* (2). The seed moisture reflects the drying system and the equilibrium of the relative humidity with the environment in which the seeds are kept.

The mean protein level of normal seeds of the accession BGH 1518 was 28.64%, whereas for BGH 7653 hull-less mutant it was 37.93% (SD ± 0.88). Such values were lower than those observed for *C. maxima* by Alfawaz (1), but higher than those obtained by Amoo *et al.* (2) and Cerqueira *et al.* (5). They were also lower than most of lines of hull-less *C. pepo* assessed by Idouraine et al. (7), but similar to those obtained by Younis et al. (14) when using African accessions cropped in areas of altitude above 2100 m.

The mean oil level of the seeds of hull-less mutant BGH 7653 was 46.70%, and higher than the mean of the seeds from the original accession BGH 1518 [41.75% (SD ± 0.59)], in accordance with Alfawaz (1) who showed an oil percentage increase after removing the hull of the seeds. Such genotypes were higher than those of *C. maxima* used by Alfawaz (1), Stevenson et al. (10) and Cerqueira et al. (5). The potential of the genotype BGH 7653 must be assessed when compared to the genotypes obtained by breeding programs of *C. pepo* for oil production. Thus, its oil concentrations are close to the best lineages selected by Idouraine et al. (7) and to the varieties used by Winkler (13) in breeding programs in Austria. Therefore, the high percentage of oil observed for hull-less mutant BGH 7653 supports the proposal regarding its usage as progenitor in breeding programs for oil extractions from seeds of *C. maxima*.

The mean level of ash obtained for the genotype BGH 7653 was 4.50% (SD ± 0.64), higher than that observed for the mean of original genotypes BGH 1518 (3.74%), shown in the Table 1. The ash levels observed for the mutant BGH 7653 were lower than those obtained for *C. maxima* by Alfawaz (1) and Cerqueira et al. (5), as well as for *C. pepo* by Idouraine et al. (7).

The mutant BGH 7653, rich in oil and protein, may increase the use of seeds as food supplement and the development of genotypes destined for food, medicinal or industrial purposes (oil extraction). The development of varieties and hull-less hybrids may result in additional income to farmer without increasing their current cropping area.

References

1. Alfawaz, M.A. 2004. Chemical composition and oil characteristics of pumpkin (*Cucurbita maxima*) seed kernels. Res. Bult. No (129): 5-18.

2. Amoo, I.A., A.F. Eleyinmi, N.O.A. Ilelaboye, and S.S. Akoja. 2004. Characterization of oil extracted from gourd (*Cucurbita maxima*) seed. Food, Agriculture & Environment 2: 38-39.

3. Andres, T.C. 2000. An overview of the oil pumpkin. Cucurbit Genet. Coop. Rept. 23: 87-88.

4. Berenji, J. 2000. Breeding, production, and utilization of oil pumpkin in Yugoslavia. Cucurbit Genet. Coop. Rept. 23: 105-109.

5. Cerqueira, P.M., M.C.J. Freitas, M. Pumar, and S.B. Santangelo. 2008. Efeito da farinha de semente de abóbora (*Cucurbita maxima*, L.) sobre o metabolismo glicídico e lipídico em ratos. Rev. Nutr. 21: 129-136.

6. Fruhwirth, G.O., and A. Hermetter. 2007. Seeds and oil of the Styrian oil pumpkin: Components and biological activities. Eur. J. Lipid Sci. Technol. 109: 1128-1140.

7. Idouraine, A., E.A. Kohlhepp, C.W. Weber, W.A. Warid, and J.J. Martinez-Tellez. 1996. Nutrient constituents from eight lines of naked seed squash (*Cucurbita pepo* L.). J. Agric. Food Chem. 44: 721-724.

8. Murkovic, M., A. Hillebrand, J. Winkler, E. Leitner, and W. Pfannhauser. 1996. Variability of fatty acid content in pumpkin seeds (*Cucurbita pepo* L.). Z. Lebensm. Unters Forsch. 203: 216-219.

9. Silva, D.J.H. da, M.C.C.L. Moura, and V.W.D. Casali. 2001. Recursos genéticos do banco de germoplasma de hortaliças da UFV: histórico e expedições. Hortic. Bras.19: 108-114.

10. Stevenson, D.G.; F.J. Eller, L. Wang, J.L. Jane, T. Wang, and G.E. Inglett. 2007. Oil tocopherol content and composition of pumpkin seed oil in 12 cultivars. J. Agric. Food Chem., 55: 4005-4013.

11. Stuart, S.G., and J.B. Loy. 1983. Comparison of testa development in normal and hull-less seeded strains of *Cucubita pepo* L. Bot. Gaz. 144: 491-500.

12. Winkler, J. 2000. Breeding of hull-less seeded pumpkins (*Cucurbita pepo*) for the use of oil. Acta Hort., 510: 123-128.

13. Winkler, J. 2000. The origin and breeding of hull-less seeded Styrian oil-pumpkin varieties in Austria. Cucurbit Genet. Coop. Rept. 23: 101-104.

14. Younis, Y.M.H., S. Ghirmay, and S.S. Al-Shihry. 2000. African *Cucurbita pepo* L.: properties of seed and variability in fatty acid composition of seed oil. Phytochemistry 54: 71-75.

Table 1. Characteristics of seed of the normal accession (BGH 1518) and hull-less mutant (BGH 7653) of *C. maxima*

Characteristic	BGH 1518	BGH 7653
Mean weight of fresh fruits (kg) (a)	2.819 (SD ± 1.54)	3.139 (SD ± 0.04)
Mean number of seeds per fruit	498.38 (SD ± 248.76)	268.04 (SD ± 51.36)
Mean weight of dry seeds per fruit (b)	78.94 (SD ± 36.42)	42.707 (SD ± 10.90)
Ratio (b/a)	2.95 (SD ± 0.94)	1.36 (SD ± 0.36)
Weight of one thousand dry seeds (g)	164.91 (SD ± 37.30)	158.34 (SD ± 10.32)
Seed length (mm)	16.8 (SD ± 1.39)	15.5 (SD ± 0.42)
Seed width (mm)	8.96 (SD ± 0.96)	9.16 (SD ± 0.62)
Seed thickness (mm)	3.13(SD ± 0.49)	2.81 (SD ± 0.16)
Humidity (%)	8.77 (SD ±1.03)	8.97 (SD ± 2.32)
Protein (%)	28.64 (nd)	37.93 (SD ± 0.88)
Oil (%)	41.75 (SD ± 0.59)	46.70 (nd)
Ash (%)	3.74 (nd)	4.50 (SD ± 0.64)

nd = not determined

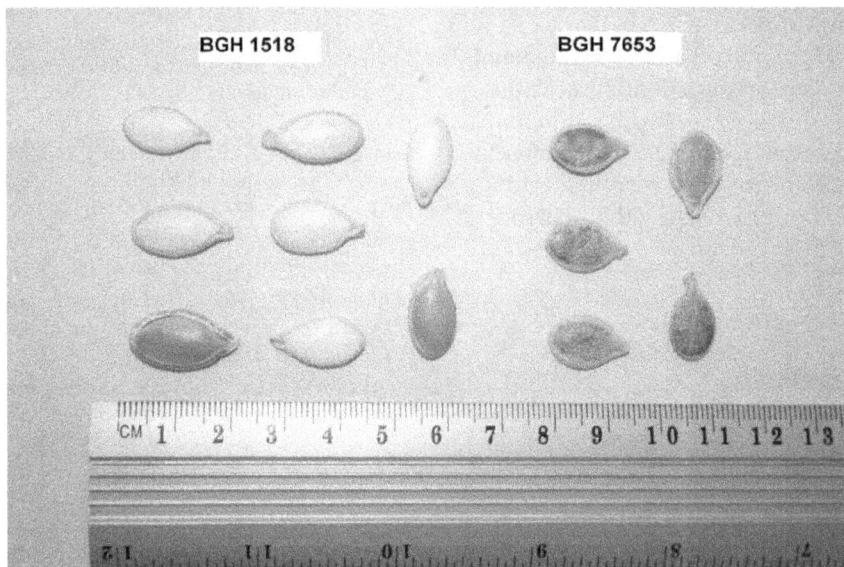

Figure 1. Morphological aspects of normal seeds of the accession BGH 1518 and mutant hull-less seeds BGH 7653.

Pollination of Squash Before and After the Day of Anthesis

R. W. Robinson

Horticulture Dept., Cornell University, Geneva, NY 14456

Cucurbit breeders usually pollinate on the morning of anthesis for both the male and the female flowers. Squash plants to be used as parents generally have male flowers open each day but may not have a female flower at anthesis on some days, preventing self pollination then. This is particularly a problem with plants grown in pots in the winter greenhouse, since they generally have fewer flowers each day than those grown in the field. It could be advantageous to be able to make pollinations before or after anthesis so that self or cross pollinations could be made on days when plants do not have open male and female flowers.

Munger (7) found that pistillate flowers of cucumber are receptive to pollination for at least 24 hours after anthesis. Pollinations he made in the winter greenhouse on the afternoon of anthesis or on the following day had good set of fruit with normal amount of seed. Wehner and Horton (10) confirmed that cucumber pollinations can be successfully made in the greenhouse on the day after anthesis.

Cucurbita pepo can also be successfully pollinated after the morning of anthesis. Summer squash pollinations I made in a winter greenhouse with open staminate flowers and with pistillate flowers on the morning after anthesis were successful.

The ability of cucurbits to produce seed from afternoon pollinations in a winter greenhouse, where the temperature can be controlled, does not guarantee that afternoon pollinations will be successful on days when high temperature may affect gamete viability. Munger (7) reported that cucumber pollinations made in the field in New York on the afternoon of anthesis were ineffective although those made in the Philippines were successful. Wehner and Horton (10) concluded that cucumber pollinations made when the temperature exceeded 35 C at North Carolina were unsuccessful.

Pistillate flowers of summer squash are receptive to pollination for at least 48 hours before anthesis. Viable seed was produced in the field and greenhouse at Geneva NY when staminate flowers at anthesis were used to pollinate pistillate flower buds that would not reach anthesis until one or two days later. Kumazama and Minikawa (6), Kodama (5), and Hayase (3) also found that *Cucurbita* stigmas are receptive to pollination on the morning of the day before anthesis. Hayase and Hiraizumi (4) reported that *Cucurbita* pollen germination was poorer on the morning of anthesis than on the previous evening but Hayase (3) and Kodoma(5) determined that pistillate flowers were more receptive to pollination on the morning of anthesis than on the previous day.

Preanthesis pollination has the advantage of eliminating the need to enclose the female flower before pollination in order to exclude pollinating insects. Another possible advantage of bud pollination is that it may be useful for difficult interspecific crosses. Hayase (1) determined that pollen tube growth was slower for unsuccessful interspecific *Cucurbita* pollinations than for successful crosses, and he reasoned that bud pollination would permit additional time for the pollen tubes to grow sufficiently to achieve fertilization.

Male as well as female gametes of *Cucurbita* are functional before anthesis. Hayase (2) and Sisa (8) reported that squash pollen had good germination on the day before anthesis. I obtained fruit with viable seed when pollen extracted with a dissecting needle from staminate flowers of *C. pepo* on the morning before anthesis was used to self pollinate open pistillate flowers.

C. pepo pollen may be viable on the day after anthesis. Some pollinations I made in the greenhouse with pistillate flowers at anthesis and staminate flowers on the day after anthesis were successful, but fruit set was less than when both flowers were at anthesis when pollinated.

Pollinating squash on weekends can be avoided without missing any pollinations by using open staminate flowers and pistillate flowers that will not be open until Saturday to pollinate on Friday. Monday pollinations can be made with staminate flowers open that day and pistillate flowers at anthesis the previous day, taking care to prevent insect pollination. Pollinations are possible with flowers opening several days apart. *C. pepo* pollinations were successfully made with male flowers on the day before anthesis and female flowers on the day after anthesis.

Pollen storage offers another opportunity to make pollinations with flowers reaching anthesis on different days. Hayase (3) stored *Cucurbita* pollen for 32 hours at 10 C, then used it to successfully pollinate pistilate flowers at anthesis. Although pollen of many species can have a long storage life if kept in a freezer at low humidity, this treatment is lethal for *Cucurbita* pollen. Wang and Robinson (9) found that freezing and low humidity both caused *Cucurbita* pollen to quickly lose viability, but pollen life was extended for as much as two weeks by storing male flower buds in a refrigerator at high humidity.

Literature Cited

1. Hayase, H. 1950. *Cucurbita* crosses. I. The pollen tube growth in interspecific crosses. Jap. J. Genet. 25: 381-390.

2. Hayase, H. 1958. *Cucurbita* crosses. XII. The fertilizing power of pollen and anther dehiscence. Jap. J. Breeding 8: 28-36.

3. Hayase, H. 1960. Physiological studies on the pollination of the genus *Cucurbita* with special reference to the fertilizing power of pollen and anther dehiscence. Hokkaido Natl. Agr. Expt. Station Rpt. 54: 1-74.

4. Hayase, H. and Y. Hiraizumi. 1955. *Cucurbita* crosses. VI. Relationship between polllen ages and the optimum conditions (saccharose concentrations and pH values) of the artificial media. Jap. J. Breeding 5: 51-60.

5. Kodama, M. 1939. On the bud pollination of *Cucurbita moschata*. Agr. Hort. 14: 1113-1115.

6. Kumazama, S. and M. Minikawa. 1937. Bud and natural polllinations in egg-plant, tomato, pepper, cucumber and squash. J. Okitsu Hort. Soc. 33: 118-130.

7. Munger, H. M. 1988. A revision of controlled pollination of cucumber. Cucurbit Genetic Coop. Rpt. 11: 8.

8. Sisa. M. 1932. Relationship between age and the viability of pollen in different cucurbits. Jap. J. Genet 8: 19-26.

9. Wang, Yong-Jian and R. W. Robinson. 1983. Influence of temperature and humidity on longevity of squash pollen. Cucurbit Genetic Coop. Rpt. 6: 91.

10. Wehner, T. C. and R. Horton Jr. 1989. Delayed pollination successful for cucumbers in North Carolina greenhouse. Cucurbit Genetic Coop. Rpt. 12:15.

Regeneration in Selected *Cucurbita* spp. Germplasm

C. Gisbert, B. Picó and F. Nuez

Instituto de Conservación y Mejora de la Agrodiversidad Valenciana (COMAV). Universitat Politècnica de Valencia, Valencia, Spain 46022

Introduction

A system for plant regeneration from individual cells or explants is essential for the application of genetic engineering. In Cucurbitaceae, *in vitro* regeneration has been reported across a wide spectrum of crops, including summer squash (*Cucurbita pepo* L.) (7), winter squash (*Cucurbita maxima* Duch.) (9), bottle gourd (*Lagenaria siceraria* Standl.) (6) and, recently, figleaf gourd (*Cuburbita ficifolia* Bouché.) (8). Genetic transformation was applied in this family for different purposes, for example to improve several cucurbits with potential to function as rootstocks (4, 6, 11). Regeneration ability is highly influenced by genotype, and then it is necessary to develop and adjust the appropriate regeneration protocols for each cultivar. In this work we assay different *in vitro* conditions useful for regeneration in eight *Cucurbita* cultivars which presented interesting characteristics for breeding. They include: three *C. pepo,* two *C. ficifolia* one *C. maxima* and two *C. moschata* species. To the best of our knowledge this is the first study reporting *in vitro* regeneration for the latter species.

Materials and Methods

C. pepo cultivars PI 171628 (a pumpkin from Turkey), V-CU-32 (a Spanish Zucchini commercial cultivar) and the commercial hybrid Temprano de Argelia (Vilmorin; TA), *C. moschata* cultivars AN-CU-45 (a Spanish Butternut cultivar) and AFR-CU-1 (a globular landrace from Morocco); *C. maxima* cultivar AN-CU-59 (a Spanish globular landrace), and the ECU-148 and Pa 06 cultivars of figleaf gourd (from Ecuador and Spain respectively) were used. The PI accession was kindly provided by NPSG-USDA Genebank and the remainder accessions, with the exception of the commercial hybrid, were maintained at the COMAV Genebank and fixed by selfing, and were morphologically and molecularly characterized by the Cucurbits Breeding group of the COMAV (1-3). Manually de-coated seeds were sequentially surface-sterilized by immersion in ethyl alcohol (96°) for 1 min, followed by 2 min in a solution of 50% commercial bleach (containing 40g L^{-1} active chloride), and 10 min in NaOCl (20%). Between each step, the seeds were rinsed once with sterile distilled water and three times

at the end of the process. Most degenerated perisperms were removed from the embryos during the sterilization procedure. The embryos were blot dried on sterile filter paper for about 3 min, and then seeds were sown in Petri dishes (5 seeds per plate) containing 25 ml of hormone-free MS medium (10) solidified with 7 g L^{-1} plant agar (Duchefa Biochemie, Haarlem, The Netherlands) and 15 g L^{-1} sucrose (Duchefa Biochemie). The pH of the medium was adjusted to 5.8 prior to autoclaving at 121°C for 20 min. The seeds were incubated at 25°C under a 16-h photoperiod with cool white light provided by fluorescent lamps (light intensity 90 µmol m^{-2} s^{-1}). Organogenic medium consists of MS salts, 30 g L^{-1} sucrose, 1mg L^{-1} 6-bencilaminopurine (BA) added after sterilization, and 7 g L^{-1} plant agar. For the accession AN-CU-45, three types of explants were used, two from cotyledons (proximal and distal from the plumule that was excised previously) and one from hypocotyls. Proximal cotyledon explants were employed in the rest of genotypes. Five explants per plate and 3 plates per genotype and treatment were used. Adventitious shoots induced from explants were isolated and cultured on MS medium or MS with 1 mg L^{-1} indole butyric acid (IBA). Results were analyzed by one-way analysis of variance and means were separated using Duncan's multiple range test.

Results and Conclusions

Explants from proximal cotyledons of *C. moschata* were found to display a markedly enhanced production of adventitious shoots compared to distal cotyledons and hypocotyls (P value <0.0000). An average of 1.33 shoots/explant was obtained from proximal explants, whereas only 0.13 were obtained from the distal and hypocotyls. These results confirm our previous results with *C. pepo* where the best performance was observed in proximal cotyledons (5) and are also similar to that recently reported in *C. ficifolia* (8). These data suggest that cells at this zone display an enhanced response to morphogenesis induction and/or that, in *Cucurbita* species, the endogenous growth regulators that promote morphogenesis accumulated at this area. Regeneration from proximal cotyledon explants was also observed in all tested genotypes on medium containing 6BA, a common growth regulator for organogenic induction (7), (Table 1). Significant differences among genotypes and cultivars belonging to similar species have been observed. *C. moschata* AFR-CU-1 had the lower regeneration response, whereas *C. pepo* cultivars PI171628 and TA showed the highest level of regeneration. The number of shoots per explant after 20 days of culture ranged from 1 to 4. Individual mean values are shown in Table 1. About 50% of isolated shoots (on average) were able to root on MB without growth regulators but, this percentage increased since 100% when IBA was added. The plantlets that were obtained were satisfactorily acclimatized. The information presented here is useful in order to select the most appropriate combination of explants/medium for regeneration in different species of *Cucurbita*. The highest regeneration ability has been found in *C. pepo*, the most economically important *Cucurbita* species. We have also reported here for the first time the regeneration ability of proximal explants in *C. moschata*, a species frequently use for breeding *C. pepo* and interesting as genetic bridge, for transferring traits from wild *Cucurbita* species into cultivated *C. pepo* and *C. maxima*.

References

1. Ferriol, M., B. Picó, and F. Nuez. 2003a. Genetic diversity of some accessions of *Cucurbita maxima* from Spain using RAPD and SBAP markers. Genetic Research and Crop Evolution. 50:227–238.

2. Ferriol, M., B. Picó, and F. Nuez. 2003b. Genetic diversity of a germ-plasm collection of *Cucurbita pepo* using SRAP and AFLP markers. Theoretical and Applied Genetics, 107:271–282.

3. Ferriol, M., B. Picó, P. Fernández de Córdova, and F Nuez. 2004. Molecular Diversity of a Germplasm Collection of Squash (*Cucurbita moschata*) Determined by SRAP and AFLP Markers. Crop Science 44:653-664.

4. Gal-On, A., D. Wolf, Y. Antignus, L. Patlis, K.H. Ryu, B.E. Min, M. Pearlsman, O. Lachman, V. Gaba, Y. Wang, Y.M. Shiboleth, J. Yang and A. Zelcer. 2005. Transgenic cucumbers harboring the 54-kDa putative gene of Cucumber fruit mottle mosaic tobacco virus are highly resistant to viral infection and protect nontransgenic scions from soil infection. Transgenic Research 14:81–93.

5. Gisbert, C., B. Picó, and F. Nuez. 2009. Regeneración de plantas a partir de explantes de cotiledón de calabacines de interés hortícola. Resúmenes VIII Reunión de la Sociedad Española de cultivo in vitro, p.51.

6. Han, J.S., C.K. Kim, S.H. Park, K.D. Hirschi, I.G. Mok. 2005. Agrobacterium- mediated transformation of bottle gourd (*Lagenaria siceraria* Standl.). Plant Cell Reports 23:692–698.

7. Kathiravan, K., G. Vengedesan, S. Singer, B. Steinitz, H. S. Paris and V. Gaba. 2006. Adventitious regeneration *in vitro* occurs across a wide spectrum of squash (*Cucurbita pepo*) genotypes. Plant Cell, Tissue and Organ Culture 85: 285–295.

8. Kyung-Min, K., K. Chang Kil, and H. Jeung-Sul. 2010. *In vitro* regeneration from cotyledon explants in figleaf gourd (*Cucurbita ficifolia* Bouche´), a rootstock for Cucurbitaceae. Plant Biotechnology Reports 4:101–107.

9. Lee, Y.K., W.I. Chung and H. Ezura. 2003. Efficient plant regeneration via organogenesis in winter squash (*Cucurbita maxima* Duch.). Plant Science 164:413–418.

10. Murashige, T., and F. Skoog. 1962. A revised medium for rapid growth and bioassays with tobacco tissue cultures. Physiologia Plantarum 15:473-497.

11. Park, S.M., J.S. Lee, S. Jegal, B.Y. Jeon, M. Jung, Y.S. Park, S.L. Han, Y.S. Shin, N.H. Her, J.H. Lee, M.Y. Lee, K.H. Ryu, S.G. Yang, and C.H. Harn. 2005. Transgenic watermelon rootstock resistant to CGMMV(Cucumber Green Mottle mosaic Virus) infection. Plant Cell Reports 24:350–356.

Table 1. Regeneration of cotyledon explants* of Cucurbita spp. on medium with 6BA mgL^{-1} after 20 days of culture

	Genotype	Percentage of explants with shoots	N° of shoots per explants
C. pepo	PI171628	100 b	2.33 d
C. pepo	V-CU-32	80 ab	1.00 ab
C. pepo	TA	80 ab	2.40 d
C. moschata	AN-CU-45	100 b	1.33 c
C. moschata	AFR-CU-1	60 a	0.80 a
C. maxima	AN-CU-59	100 b	1.13 ab
C. ficifolia	ECU-148	80 ab	0.80 a
C. ficifolia	Pascual 06	67 a	0.80 a

*Cotyledons containing the region proximal to the plumule which has been previously excised.
Means followed by different letter are significantly different by Duncan's multiple comparison test.

"Exploding" Fruits not Unique to Watermelon: Fruit Cracking in *Cucurbita moschata*

Linda Wessel-Beaver

Dept. of Crops & Agroenvironmental Sciences, University of Puerto Rico, PO Box 9000, Mayaguez, Puerto Rico, USA (lindawessel.beaver@upr.edu)

In May 2011, a Chinese TV station carried reports of "exploding" watermelons in Danyang in Jiangsu province in eastern China. This report was picked up by a number of internet news services (4,7,12). The use of the growth regulator forchlorfenuron too late in the season and under rainy conditions was cited as the cause.

Ruptures occurring in either the rind or cuticle are reported in a number of fruit including tomato (14), cherry (9,10), grape (3), pepper (2,5) and noni (*Morinda citrifolia*) (6). Two types of cracking are generally mentioned: radial (beginning at the stem) and concentric (circular). In pepper, small fissures (russeting) can also occur (5). In the case of the "exploding" Chinese watermelons, both radial and concentric cracking are seen in photographs posted to the web (4,7,12).

Most researchers attribute fruit cracking to the amount of water available to the plant as the fruit is ripening (1, 3, 5, 6, 9, 10, 11). Cracking appears to be the result of the buildup of internal turgor pressure caused by excess water availability. It may be especially common when insufficient water availability is followed by excess irrigation or rainfall. In pepper, Aloni et al. (2) proposed that the nighttime reduction in transpiration results in the high turgor pressure that causes cracking in peppers. A number of authors mention that the incidence of fruit cracking can vary among cultivars (1, 3, 9, 10, 11).

One of the goals of the tropical pumpkin (*Cucurbita moschata*) breeding program in Puerto Rico is to develop semi-bush cultivars with high dry matter content and deep orange color. The genetic materials used in this program derive from crosses between temperate (primarily 'Bush Butternut') and tropical material (13). We have often noticed that some of the derived lines are susceptible to fruit cracking. 'Taína Dorada', a cultivar released from this program, is appreciated by growers for its fruit quality, especially its deep orange color and high dry matter content. Percent dry matter and °Brix in 'Taína Dorada' is much higher than that of the Puerto Rico standard cultivar 'Soler'. Despite these positive attributes, growers complain that they regularly encounter cracked fruit. Both radial and concentric cracks occur (Figure 1).

This problem has never been observed in 'Soler', even when both varieties have been planted in fields where irrigation management and rainfall is the same. Cracking usually occurs just as the fruit is maturing (>30 days post-anthesis) and can be up to 3 cm deep. The fruit cavity is not affect, the exposed flesh quickly suberizes, and fruit rotting does not occur. However, these fruits are no longer marketable.

Splitting or cracking of squash fruit, particularly butternut types, is not an uncommon problem judging by the number of home gardeners sending questions or comments to internet gardening forums or blogs (for examples, see http://www.no-dig-vegetablegarden.com/butternut-squash-splitting.html and http://www.idigmygarden.com/forums/archive/index.php/t-10379.html). According to a technical bulletin for Sakata Seed (1) all squash fruit have the potential to crack and susceptibility varies among cultivars. 'Taína Dorada' has both the thin rind and high sugar content that, according to this bulletin, make a variety more susceptible to fruit cracking. High sugar levels result in a higher osmotic potential in the fruit, thus favoring greater uptake of water. The result is that fruit cells swell causing the fruit to crack. The bulletin notes that mature fruit are particularly affected and that over fertilization may play a role.

Strang et al. (11) evaluated 24 squash and pumpkin varieties in Kentucky during a very wet season. Several cultivars had cracked fruit, including one that continued to crack when exposed to rain after harvesting. Included in this trial was the tropical pumpkin cultivar 'La Estrella' which is very closely related to 'Taína Dorada'. However, no fruit cracking was reported in 'La Estrella' in this trial.

Selection against fruit cracking might not be effective if, at the same time, one wishes to select for increased fruit sugar and dry matter content. The best approach for minimizing this problem is likely to be good water management and prompt harvesting of mature fruit.

Literature Cited:

1. Anonymous. 2008. Sakata Technical Bulletin, Atlas Hybrid F1 Squash. Retrieved online 14 June 2011: <http://www.mayford.co.za/VegSeedQrevDocs/Squash%20Butternut/Commercial%20Varieties/ATLAS.pdf>

2. Aloni, B., L. Karni, I. Rylski, Y. Cohen, Y. Lee, M. Fuchs, S. Moreshet and C. Yao. 1998. Cuticular cracking of pepper fruit. I. Effects of night temperature and humidity. J. Horti. Sci & Biotechnology 73(6):743-749.

3. Beede, R.H. No date. Berry cracking in table grapes. Retrieved 4 June 2011: <http://cekings.ucdavis.edu/files/18991.pdf>

4. Foster, P. 2011. China goes organic after years of 'glow in the dark pork' and 'exploding watermelons.' Telegraph, May 29, 2011. Retrieved online 4 June 2011: <http://www.telegraph.co.uk/news/worldnews/asia/china/8544851/China-goes-organic-after-years-of-glow-in-the-dark-pork-and-exploding-watermelons.html>

5. Jovicich, E., D.J. Cantliffe, S.A. Sargent, and L.S. Osborne. 2004. Production of greenhouse grown peppers in Florida. Retrieved 14 June 2011: < http://edis.ifas.ufl.edu/hs228>

6. Nelson, S.C., 2006, December 7. The noni website. College of Tropical Agriculture and Human Resources, Univ. of Hawaii at Mânoa. Retrieved 1 June 2011: <http://www.ctahr.hawaii.edu/noni/cracking.asp>

7. Olesen, A. 2011. Fields of watermelon burst in China farm fiasco. Associated Press. May 17, 2011. Retrieved online 4 June 2011: <http://news.yahoo.com/s/ap/20110517/ap_on_fe_st/as_china_exploding_watermelons>

8. Sekse, L. Fruit cracking mechanisms in sweet cherries (*Prunus avium* L.) – a review. Acta Hort. (ISHS) 468:637-648. Retrieved online 5 June 2011: <http://www.actahort.org/books/468/468_80.htm>

9. Simon, G. 2003. Az ültetvények védelme az esô és a madarak által okozott károk ellen. In rotkó, K. (szerk): Cseresnye és meggy. Mezôgazda Kiadó, Budapest, 338–348. [cited in Simon, 2006].

10. Simon, G. 2006. Review on rain induced fruit cracking of sweet cherries (*Prunus avium* L.), its causes and the possibilities of prevention. International Journal of Horticultural Science. 12 (3): 27–35 (Retrieved online 4 June 2011: <http://www.agrarkutatas.net/files/aktualis/pdf_agroinform_20070220124757_05_Simon_05.p>

11. Strang, J., K. Bale, J. Snyder and C. Smigell. 2007. Winter squash and pumpkin variety evaluations Retrieved online 14 June 2011: <http://www.hort.purdue.edu/fruitveg/rep_pres/2006-7/mvvt_2006_pdf/Winter_SquashPurdCvTr.pdf>

12. Watts, J. 2011, 17 May. Exploding watermelons put spotlight on Chinese farming practices. Retrieved 17 May 2011: <http://www.guardian.co.uk/world/2011/may/17/exploding-watermelons-chinese-farming>

13. Wessel-Beaver, L., O. Román-Hernández and L.E. Flores-López. 2006. Performance of new tropical pumpkin genotypes under varying cultural practices. J. Agric. Univ. P.R. 90(3-4):193-206.

14. White, D.G. and Whatley, B.T. 1955. A method to measure cracked apples and tomatoes. Proc. Amer. Soc. Hort. Sci. 65:289-290.

Figure 1: Fruit cracking in *Cucurbita moschata* cv. Taína Dorada. This planting, in Mayagüez, Puerto Rico, was drip irrigated; however, extremely heavy rains occurred late in the season. Cracks are up to 2.5 cm deep. The variety 'Soler' planted in the same field did not show fruit cracking.

Occurrence and Preliminary Characterization of Gynoecious Ridge Gourd [*Luffa acutangula* (L.) Roxb.] in a Natural Population

A. D. Munshi, T. K. Behera, A. K. Sureja and Ravinder Kumar

Division of Vegetable Science, Indian Agricultural Research Institute, New Delhi-110012

A gynoecious plant of ridge gourd (*Luffa acutangula*) with absolute expression of gynoecism collected from Hoogly district of West Bengal, India, was identified and characterised at Indian Agricultural Research Institute, New Delhi. This is the first report of naturally occurring gynoecism in *Luffa acutangula*. Efforts were made to maintain the line by crossing with the sister line. With its preliminary form, few crosses were also attempted with released cultivars of ridge gourd 'Pusa Nasdar' and 'Pusa Nutan' to examine the possible involvement of genetic/ environmental factors in the expression of gynoecious flowering habit.

Ridge, angled gourd, or angled loofah (*Luffa acutangula* (L.) Roxb.) is a cucurbitaceous vegetable originated in sub-tropical region of Asia. India is considered as a primary centre of origin (Chakravarty, 1990). This crop is cultivated in India, Southeast Asia, China, Japan, Egypt and other parts of Africa. Ridge gourd is generally monoecious in nature with pistillate (female) flowers borne in axil of flowers and staminate (male) flowers in raceme (Fig. 1). Moreover, it has an ancestral form "Satputia" found in Bihar (one of the eastern states of India) which is hermaphrodite (Fig. 1; both male and female organs in same flower) in nature and was given a separate taxonomy status as *Luffa hermaphrodita* (Singh and Bhandari, 1963). Monoecious ridge gourd is late in maturity and produces large size fruit with less number of fruits. Whereas, hermaphrodite form is early in maturity and produces large number of small size fruits in clusters.

The occurrence of gynoecism in *Luffa acutangula* is very rare. A gynoecious plant (only female flower in a plant) of *Luffa acutangula* with absolute expression of gynoecism, which was collected from Hoogly district of West Bengal (another eastern state of India) was identified and characterised at Indian Agricultural Research Institute, New Delhi. This is the first report of naturally occurring gynoecism in *Luffa acutangula*, though male sterility in ridge gourd was reported earlier by Pradipkumar *et al.* (2007). In this paper attempt has been made to present the preliminary observations of plant and fruit characters of the gynoecious plants.

The potential use of gynoecy in increasing cucumber yield was studied by several workers. The gynoecious sex form in bitter gourd (*Momordica charantia* L.) has been reported in recent past from India (Behera *et al.*, 2006) which is under the control of a single recessive gene (Behera *et al.*, 2009). It was found that the number of female flowers was positively correlated with yield in some population-season combinations. Moderate to highly significant positive correlations (r) between per cent pistillate nodes and yield were also identified in cucumber suggesting sex expression has potential for increasing yield through indirect selection (Cramer and Wehner, 2000). A positive correlation ($r = 0.24$ to 0.40) was observed with the number of females nodes on lateral branches and total fruit per plant (Fan *et al.*, 2006). Like that of cucumber and bitter gourd, the gynoecism in *Luffa* reported in this study will be helpful for significant increase in yield through easy and economic way of hybrid development and hybrid seed production in this crop.

Salient features of gynoecious *Luffa acutangula*

1. Plant type: Trailing or climbing vines with angular stem and tendrils with vines of 1.5 -2.0 m in length.
2. Leaf characters: Leaves dark green, 3-5 lobed, shallow lobbing, nearly glabrous, smooth with long petioles. Internode length 7.0-7.5 cm.
3. Flower characters: Gynoecious, petals yellow and showy, female flower solitary (Fig. 1) in long pedicel (5.0-6.5 cm), occasionally in clusters, ovary long (3.5-5.0 cm), slightly ribbed stigma, trifid, anthesis time 5.30 pm – 6.30 pm.
4. Fruit characters: Fruits long (15-20 cm), straight, light green, smooth with very shallow ridges. Average fruit weight 55-65 g, tender flesh.
5. Disease reaction: Field tolerance to *Luffa yellow mosaic virus*.
6. Maturity: Ready for first harvesting 55-60 days.

   Since the gynoecious plant was identified at latter stage, i.e. at flowering stage, therefore spraying of growth regulators like GA_3 and silver thiosulphate did not have any effect for induction of male flowers for its maintenance. Efforts were made to maintain the line by sibbing with one of the sister line and simultaneously crossing was also attempted with released cultivars of ridge gourd 'Pusa Nasdar' and 'Pusa Nutan' to examine the possible involvement of genetic/ environmental factors in the expression of gynoecious flowering habit. Vine cutting and tissue culture are also being attempted to maintain the line for its further exploitation.

Literature Cited

Behera, T.K., S.S. Dey and P.S. Sirohi. 2006. 'DBGy-201' and 'DBGy-202':two gynoecious lines in bitter gourd (*Momordica charantia* L.) isolated from indigenous source. Indian Journal of Genet 66: 61–62.

Behera, T.K., S.S. Dey, A.D. Munshi, A.B. Gaikwad, Anand Pal, and Iqbal Singh. 2009. Sex inheritance and development of gynoecious hybrids in bitter gourd (*Momordica charantia* L.). Scientia Horticulturae 120: 130–133.

Chakravarty, H.H. 1990. Cucurbits in India and their role in development of vegetable crops. pp 325–348. In: Bates, D. M., Robinson, R. W. and Jeffrey, C. (eds.), Biology and Utilization of the Cucurbitaceae. Cornell University Press, Ithaca, New York.

Cramer, C.S. and T.C. Wehner. 2000. Path analysis of the correlation between fruit number and plant traits of cucumber populations. HortScience 35: 708–711

Fan, Z., M.D. Robbins and J.E. Staub. 2006. Population development by phenotypic selection with subsequent marker-assisted selection for line extraction in cucumber (*Cucumis sativus* L.). Theoritical and Applied Genetics 112: 843–855.

Pradeepkumar, T., R. Sujatha, B.T. Krishnaprasad and I. Johnkutty. 2007. New source of male sterility in ridge gourd (*Luffa acutangula* (L.) Roxb.) and its maintenance through in vitro culture. Cucurbit Genetics Cooperative Report 30: 60–63.

Singh, D. and M.M. Bhandari. 1963. The identity of an imperfectly known hermaphrodite *Luffa*, with a note on related species. *Baileya* 11: 132–141.

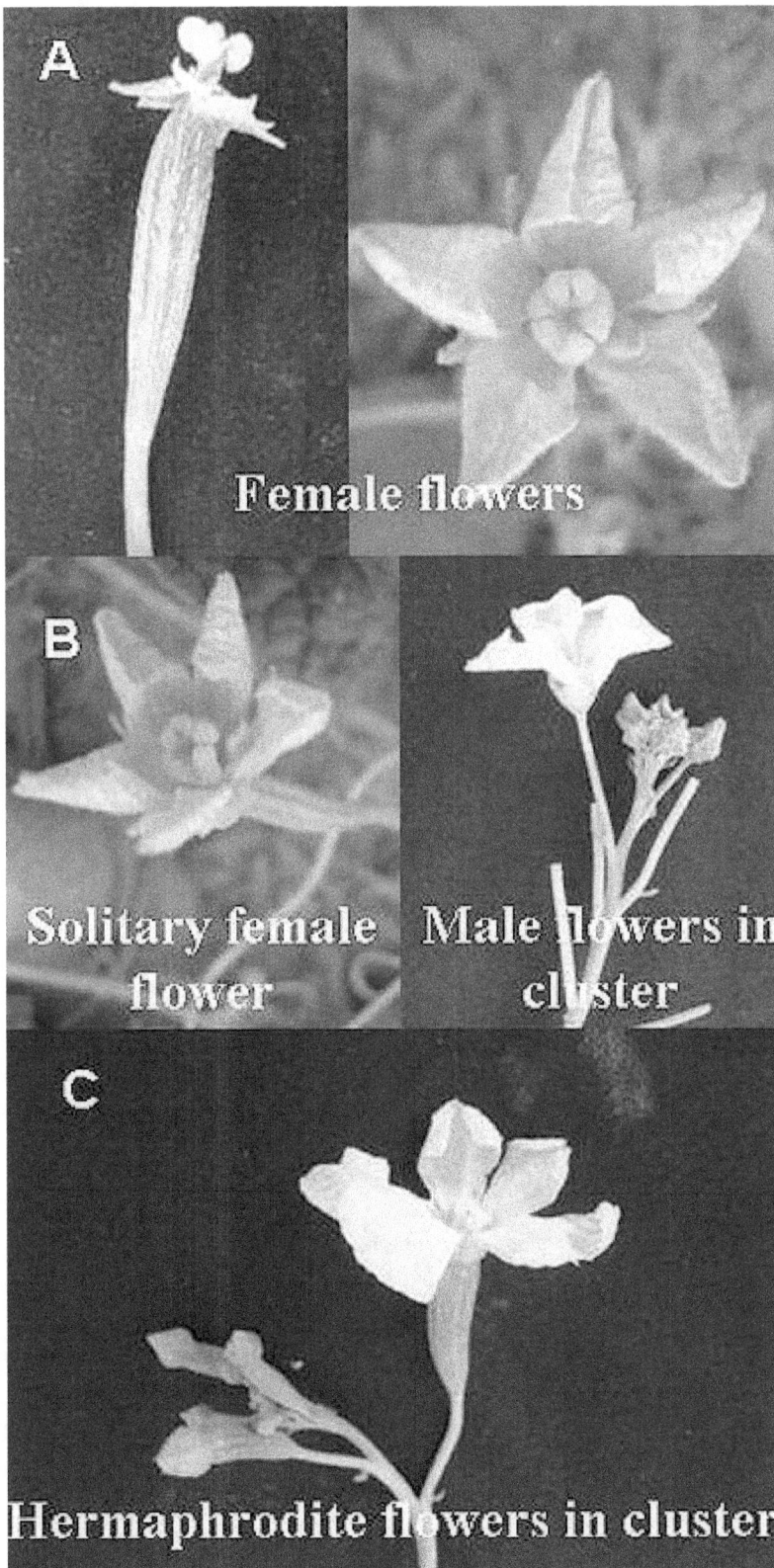

Figure. 1 A. Female flower of gynoecious *Luffa acutangula*; B. Female flower and male flowers of monoecious *L. acutangula*; C. Hermaphrodite flowers of *L. hermaphrodita*

Use of Silver Thiosulfate and Gibberellic Acid for Induction of Hermaphrodite Flower in Gynoecious Lines of Bitter Gourd (*Momordica charantia* L.)

T. K. Behera, Smaranika Mishra and Anand Pal
Division of Vegetable Science, Indian Agricultural Research Institute, New Delhi-110012

Bitter gourd is considered as one of the most nutritious gourds and grown for its fruits and leaves which have medicinal properties. The genetic manipulation which has shown potential for increasing sex stability and uniform flowering is the chracterization and introduction of the hermaphroditic character (1). Cucumber hybrids made using hermaphrodites have been hypothesized to be sexually more stable under conditions of environmental stress (2).The predominant sex form in bitter gourd is monoecious, however, gynoecious sex form (only female flower in a plant) has been reported in two spontaneous seedling population (DBGy 201 and DBGy 202) at Indian Agricultural Research Institute, India (3). At the beginning these two lines have been maintained using sib-mating which resulted into very few gynoecious plants in subsequent generations. With the help of in vitro techniques a large number of plants has been generated for sex modification study using DBGy 201 genotype. The present study was undertaken to determine chemicals threshold concentration and dosage of GA_3 Silver Nitrate and Silver Thiosulfate which would cause gynoecious genotypes to produce hermaphrodite flowers.

Recommended cultural practices and plant protection schedule were followed. Silver Thiosulphate at 6M and 3M, Silver Nitrate at 250ppm and 200 ppm and GA_3 at 1500 ppm and 1000 ppm were treated at after appearance of 1st female flower. Treatments were arranged in a randomized complete block design with 4 replications having 2 subsamples. Plants were allowed to grow throughout the life cycle.

Data were observed on node at which 1st female flower appear, ovary length of female flower (cm) and hermaphrodite flower (cm) and number of female flower and hermaphrodite flower per plant during the growing period. The results depicted in table 1 showed the effectiveness of Silver Thiosulfate 6M for sex modification in bitter gourd. Silver nitrate did not show any response in sex modification in both concentrations. It was found that there was comparatively less fruit set in the hermaphrodite flower compared with the female flower those were pollinated with pollen of hermaphrodite flowers of same plant (Fig. 1). This may be because of the splitting of stigmatic surface of hermaphrodite flowers. Hence, further experimentations are in progress to find out threshold concentration and dosage of Silver Thiosulfate for obtaining functional hermaphrodite flowers in gynoecious genotype.

References

1. Kubicki, B. 1965. New possibilities of applying different sex types in breeding. Genet. Pol. 6:241-249.

2. Pike, L.M. and W.A. Mulkey. 1971. Use of hermaphrodite cucumber lines in the development of gynoecious hybrids. HortScience 6:339-340.

3. Behera, T. K., S. S. Dey, and P. S. Sirohi. 2006. DBGy-201 and DBGy-202: two gynoecious lines in bitter gourd (*Momordica charantia* L.) isolated from indigenous source. Indian J Genet 66: 61-62.

Table 1. Conversion of female flower into hermaphrodite flower in gynoecious genotype of bitter gourd, DBGy 201.

Treatment	Node at which 1st female flower appear	Ovary length of female flower (cm)	Ovary length of hermaphrodite flower (cm)	Number of female flower/ plant	Number of hermaphrodite flower/ plant
Silver Thiosulfate 6M	7.13	1.31	3.02	70.22	95.60
Silver Thiosulfate 3M	7.46	1.18	3.11	103.34	79.70
Silver Nitrate 250ppm	7.6	1.35	0	54.45	0
Silver Nitrate 200 ppm	7.33	1.34	0	133.11	0
GA$_3$ 1500 ppm	7.26	1.24	2.53	77.22	12.00
GA$_3$ 1000 ppm	6.93	1.38	2.67	60.34	13.90

A. Female flower   B. Initiation of stamens   C. Hermaphrodite flower

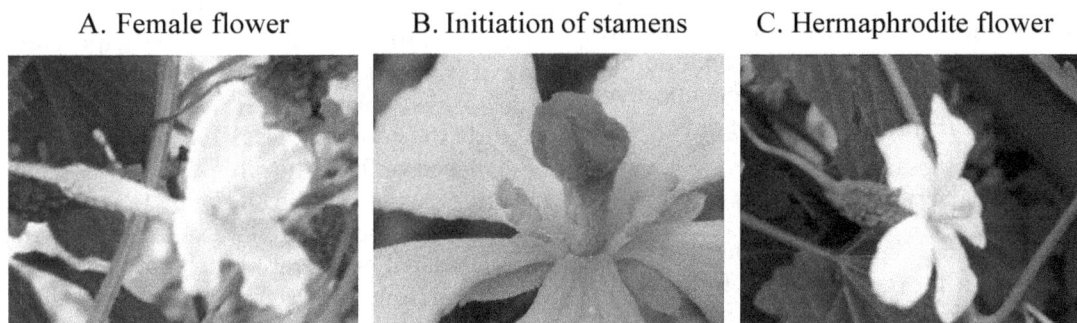

Figure 1. Sex conversion in gynoecious bitter gourd genotype by using Silver Thiosulfate 6M.

Phenotypic Diversity Analysis in Pointed Gourd (*Trichosanthes dioica* Roxb.)

L. K. Bharathi and Vishalnath

Central Horticultural Experiment Station (IIHR), Aiginia, Bhubaneswar-751019, Orissa, India

Introduction

Pointed gourd (*Trichosanthes dioica* Roxb.) is one of the most nutritive cucurbit vegetables and is native to Indian subcontinent. It is one of the important vegetables of India and Bangladesh. Pointed gourd is rich in vitamin and contains 9.0 mg Mg, 2.6 mg Na, 83.0 mg K, 1.1 mg Cu, and 17.0 mg S per 100 g edible part. It is reported to have the medicinal property of lowering blood sugar and total cholesterol (1, 2 & 3) and also known to have antiulcerous effects (5). Knowledge on magnitude and nature of genetic variation helps in formulating breeding programme for improvement of a crop. The objective of the present study was to use a diverse set of pointed gourd germplasm to estimate the extent of genetic diversity with respect to yield and yield components.

Materials and Methods

A total of 22 varieties/land races of pointed gourd were collected from different states of India (Uttar Pradesh, Bihar and Orissa) and were planted at a spacing of 2 × 1m in a randomized block design with three replications during 2008-2010. The plant population in the experimental field had female and male plants at the ratio of 10:1 to ensure effective pollination. The accessions were evaluated for seven morphological characters. Mean, ANOVA, and coefficient of variation were analyzed by using the AgRes statistical software (version 7.1) and mean data was further subjected to correlation analysis by using GenRes statistical software (version 7.01).

Results and Discussion

The analysis of variance revealed significant differences among the accessions studied for all the characters. A wide range of variation (Table 1) was observed for internode length (5.50–13.67 cm), node number bearing first flower (6.00– 16.83), number of fruits per plant (79.86–502.00), fruit weight (14.31-37.98 g), fruit length (4.17-8.73 cm) and yield per plant (2.48-10.21 kg). Among the accessions studied, the earliest flowering was observed in CHPG 11 (51 d), which is a favourable trait to

take advantage of early harvest and profitable market avenue owing to its high demand. The variety Swarna Alaukik recorded the highest yield (10.21 kg/plant) followed by CHPG 11 (9.57 kg/plant).

The accessions showed a considerable level of variability for qualitative traits such as fruit colour (Fig. 1; light green to dark green), fruit striping (Fig.1; striped to without striped), fruit shape (round to oblong), leaf surface (smooth to rough). The high level of phenotypic diversity observed among the accessions in this study is in agreement with the earlier findings (4) for different characters in other sets of germplasm. The diversity found in fruit colour and shape would be helpful for the selection of genotypes with desired quality. In addition, the accessions displaying contrast characters can be used to study the genetics of trait expression.

Based on D^2 values the accessions were grouped in to five clusters (Table 2). The pattern of group constellations indicated that genetic diversity was not directly correlated to the geographic diversity. Cluster I was largest comprising 12 accessions, whereas cluster V was smallest with a single accession. The intra-cluster distance (Table 3) ranged from 0.00 (cluster V) to 26.67 (cluster I) and inter-cluster distances varied from 17.60 (cluster III Vs cluster V) to 47.76 (cluster IV Vs cluster V). Cluster IV could be characterized (Table 4) with genotypes early in flowering (<55 days), higher values for fruit characters (fruit weight-32.78 g; fruit length-8.72 cm) and yield per plant (9.59 kg). Though maximum number of fruits per plant (466.01) was recorded in cluster II it stood second in terms of yield per plant which may be due to less fruit weight recorded by this group. The results presented here on diversity existing in this species will provide the foundation for designing an efficient pointed gourd breeding programme.

Literature cited

1. Chandrasekar, B., B. Mukherjee, and S.K. Mukherjee. 1988. Blood sugar lowering effect of *Trichosanthes dioica* Roxb. in experimental rat models. Int. J. Crude Drug Res 26:102-106.

2. Ram D. 2001. Non-hierarchial Euclidean cluster analysis in pointed gourd (*Trichosanthes dioica* Roxb.). Indian J Hort 58: 2001

3. Sharma, G., and M.C. Pant. 1988. Effects of feeding *Trichosanthes dioica* (parval) on blood glucose, serum triglyceride, phospholipid, cholesterol, and high density lipoprotein-cholesterol levels in the normal albino rabbit. Current Sci 57:1085–1087.

4. Sharma, G., M.C. Pant, and G. Sharma. 1988. Preliminary observations on serum biochemical parameters of albino rab-bits fed on *Trichosanthes dioica* (Roxb.). Indian J Medical Res 87:398–400.

5. Som, M.G., T.K. Maity, and P. Hazra. 1993. Pointed Gourd. in Genetic Improvement of Vegetable Crops edited by Kalloo G and Berg BO, Pergamon Press, Oxford, UK

Table 1. *Per se* performance of 20 Accessions for different traits of pointed gourd

Accession	Days to flowering	Node no bearing 1st flower	Internodal length (cm)	Fruit weight (g)	Fruit length (cm)	No of fruits/plant	Yield/plant (kg)
CHPG 1	69.33	12.16	6.83	24.15	7.11	127.86	2.95
CHPG 2	73.50	15.17	9.33	22.69	6	131.67	3.59
CHPG 4	70.67	9.67	9.17	32.34	7.83	205.78	6.06
CHPG 7	67.72	16.77	7.50	32.85	6.7	201.42	6.07
CHPG 8	53.22	13.61	11.17	23.89	6.93	238.25	6.44
CHPG 11	70.83	12.50	10.33	35.84	7.83	296.78	9.57
CHPG 12	63.22	12.44	9.33	37.98	8.27	215.50	6.68
CHPG 14	63.33	9.83	7.67	37.68	7.87	79.86	2.48
CHPG 15	47.17	12.44	8.17	37.90	7.6	109.83	3.73
CHPG 16	60.50	12.11	11.33	24.72	6.67	95.25	2.65
NP 309	65.66	15.11	10.33	20.51	6.17	239.08	7.07
NP 702	54.67	7.33	8.17	23.26	6.67	484.53	9.73
NP 709	63.88	10.66	10.50	15.70	5	462.67	6.43
NP 751	50.83	6.00	11.00	23.47	5.23	502.00	9.43
NP 754	57.67	7.00	6.33	27.50	6.73	414.83	8.26
NP 756	62.55	10.56	10.17	29.52	7.57	244.92	6.67
NP 308	70.00	11.11	5.50	16.00	4.17	181.67	2.89
Adauri Hara	61.00	7.78	9.83	29.61	8.57	143.58	4.70
Rajendra Parwal	51.44	7.44	7.67	31.36	8.73	308.33	8.97
Swarna Alaukik	56.67	10.72	13.67	34.20	8.7	374.33	10.21
HAP 39	66.00	15.27	13.67	31.11	8.03	154.33	4.76
HAP 28	57.00	16.83	6.67	14.31	4.67	152.33	3.78
Mean ± SE	61.68±1.69	11.48±1.21	9.27±0.58	27.57±1.25	6.96±0.26	243.86±14.10	6.05±0.12
CD (P=0.05)	3.41	2.45	1.17	2.51	0.51	28.46	0.24

Table 2. Composition of clusters based on genetic divergence in pointed gourd

Cluster	No. of genotypes	Genotypes and their source
I	12	Orissa- CHPG 1, CHPG 2, CHPG 4, CHPG 7, CHPG 8, CHPG 11, CHPG 12, CHPG 14, CHPG 15, CHPG 16 Uttar Pradesh- NP 309, NP 756
II	4	Uttar Pradesh-NP 702, NP 709, NP 751, NP-754
III	3	Uttar Pradesh-NP 308, Adauri Hara Bihar-HAP-39
IV	2	Uttar Pradesh-Rajendra Parwal Bihar-Swarna Alaukik
V	1	Bihar-HAP-28

Table 3. Inter- and intra- cluster distances

Cluster	I	II	III	IV	V
I	26.67	34.15	24.05	38.99	25.24
II		19.92	37.61	24.36	37.02
III			21.44	45.51	17.60
IV				15.05	47.76
V					0.00

Table 4: Cluster mean for different traits of pointed gourd

Cluster	Days to flowering	Node no bearing 1st flower	Internodal length (cm)	Fruit weight (g)	Fruit length (cm)	No of fruits/plant	Yield/plant (kg)
I	63.98	12.70	9.28	30.01	7.21	182.18	5.33
II	56.76	7.75	9.00	22.48	5.91	466.01	8.46
III	65.67	11.39	9.50	25.57	6.92	159.86	4.12
IV	54.06	9.08	10.67	32.78	8.72	341.33	9.59
V	57.00	16.83	6.67	14.31	4.67	152.33	3.79

Clonal Selection-NP 702

cv. Swarna Alaukik

Figure 1. Fruits of NP 702 and Swarna Alaukik

Performance of Gynoecious × Monoecious Hybrids of Bitter Gourd (*Momordica charantia* L.)

Swati Khan and T. K. Behera

Division of Vegetable Science, Indian Agricultural Research Institute, New Delhi-110012

Bitter gourd (bitter melon; *Momordica charantia*) is an important cucurbitaceous vegetable crop grown in the tropics. It has rich nutritional and medicinal values. The monoecious bitter gourd accessions produce staminate flowers from the start of reproductive phase till crop maturity (1) and thus the staminate to pistillate flower sex ratio in this sex type is relatively high (9:1 to 48:1; 3). Moreover, it creates difficulty during commercial hybrid seed production due to its extremely small flower. Use of gynoecious line is an alternative to reduce the cost of hybrid seed production. Gynoecious lines evolve spontaneously in seedling population and once isolated can be exploited for hybrid development by eliminating hand pollination. Two gynoecious lines (DBGy 201 and DBGy 202) lines have been developed from natural population at IARI (1). Development of hybrids in bitter melon is expensive because of hand pollination but the utilisation of gynoecy is economical and easier for exploiting hybrid vigour in bitter gourd (2). The present study was undertaken to compare the per se performance of gynoecious × monoecious hybrids with monoecious × monoecious hybrids in bitter gourd.

The experiment was conducted to evaluate the performance of hybrids in bitter melon by making crosses on the tissue cultured raised gynoecious line, DBGY-201 with twelve other inbreds. The twelve different inbreds (Pusa Do Mausami, Pusa Vishesh, Sel 2, Nakhara Local, MC 84, S26, S29, S30, S41, S54, S57 and DBG 34) were crossed with DBGY-201 to produce 12 crosses. The F_1 hybrids along with three commercial monoecious × monoecious hybrids namely VNR 22, US 33, PH 2 were used as checks to compare the performance of gynoecious × monoecious hybrids. During wet season (July to September) season of 2010, these 12 hybrids along with the 3 checks were grown at Experimental Farm, Division of Vegetable Science, Indian Agricultural Research Institute, New Delhi, India. The experiment was laid out in randomised block design with 3 replications. The seedlings were transplanted on both sides of the channel with 2m between channel and 45 cm between plants. The recommended NPK fertilizer doses and cultural practices along with plant protection measures were followed to raise an ideal crop. The bitter gourd is mainly grown in long growing spring-summer season (February to May) in north Indian plains, but this experiment was carried out in wet season in order to evaluate the performance of these hybrids in off season. Because of the short period growing season i.e. July-September only two harvests were taken.

Per se performance of hybrids

Time of harvest is a critical consideration for maximizing financial gain in commercial crops. Earliness in bitter gourd is attributed to node number to first female flower and time required for first female flower appearance. In the present findings it was found that gynoecious × monoecious hybrids, DBGy- 201 x S54 had first female flower at 3rd node followed by DBGy- 201 x DBG 34 at 5thnode whereas monoecious × monoecious hybrids like VNR 22 showed first female flower at 11th node and Pusa Hybrid 2 at 9th node (data not presented). Minimum days required to first fruit harvest also depict the earliness of the crops. Minimum numbers of days were taken by DBGY- 201 x S54 with 41 days followed by DBGY- 201 x DBG 34 with 42 days whereas among the checks Pusa Hybrid 2 (monoecious × monoecious) required more days (48 days) for first fruit harvest and VNR 22 with maximum number of days of 51. Early picking is desirable to catch the early market for better remuneration and provides ample scope for crop rotation with other crops in the same field. This might have resulted due to the transfer of earliness character from the gynoecious parent to the hybrid. Like cucumber, earliness in bitter gourd is judged through appearance of female flower at lower node and days required for first picking.

Sex ratio determines fruitfulness of most crops in general and cucurbitaceous crops in particular (5). The higher the proportion of female flower greater is the productivity. Lowest sex ratio (male:female) was observed in DBGy- 201 x S26 (1.56) followed by DBGy- 201 x S29 (2.07) whereas highest sex ratio was found in DBGy- 201 x MC 84 (5.15) among all gynoecious × monoecious hybrids and among all hybrids PH2 (check) showed sex ratio as high as 10.22. DBGy- 201 x S54 was found superior than the other hybrids and even checks as the number of fruits per plant was maximum in this hybrid (12.79) followed by DBGy- 201 x DBG 34 with 8.85 fruits/plant and hybrid PH2 produced only 3.55 fruits. Numbers of fruits and fruit weight directly determine bitter gourd yield (3). In cucumber, Fan et al. (4) identified a positive correlation between yield and sex expression. Hence, these aforesaid gynoecious hybrids of bitter gourd have high potential for commercial cultivation with respect to yield and earliness.

Fruit length and diameter are also important attributes to determine yield. The maximum fruit length was in DBGy- 201 x S54 with 11.07 cm followed by DBGy- 201 x Pusa Vishesh with 10.87 cm (Table 1). The maximum fruit diameter was registered in DBGy- 201 x Pusa Vishesh (11.87 cm) and DBGy- 201 x Nakhara Local (11.30 cm). The parents with the larger fruits might have contributed to increase the fruit length in hybrids. Among the checks VNR 22 recorded the maximum fruit diameter (12.20 cm). The maximum fruit weight was registered in hybrid Gy x S54 (39.18 g) and was at par with the check VNR 22 (39.43 g). DBGy- 201 x Pusa Vishesh hybrid had also larger fruit weight (38.43 g) while the minimum was in DBGy- 201 x S29 (23.77 g) and the monoecious × monoecious hybrid US 33 produced smallest fruit (15.07 g). Maximum yield per plant (245.86 g) was registered in hybrid DBGy- 201 x Pusa Vishesh with followed by DBGy- 201 x MC 84 (240.21 g). It was found that all the hybrids performed extremely well in comparison to checks in terms of yield. The gynoecious parent holds immense potentiality for exploitation of hybrid vigour with respect to yield and earliness (2). They also reported that DBGy-201 is a good combiner and it exhibited significantly GCA and SCA effects with most of the parents.

The gynoecious line DBGy- 201 taken in the present study could be utilized for heterosis breeding for high yield and earliness. Besides this, it has been proved that this line has already shown high general combining ability (GCA), hence, selection in advanced generations is also equally awarding for development of predominantly gynoecious cultivars by using DBGy-201 as female parent in bitter gourd (2). The potentiality of gynoecious lines in term of yield and earliness was well recognized in cucumber and also in this present study on bitter gourd. Thus, from the above discussion it become clear that gynoecious parent incorporated in this study holds immense potentiality for utilization in hybrid bitter gourd.

References

1. Behera, T. K., S. S. Dey, and P. S. Sirohi. 2006. DBGy-201 and DBGy-202: two gynoecious lines in bitter gourd (*Momordica charantia* L.) isolated from indigenous source. Indian J Genet 66: 61-62.

2. Dey, S. S., T. K. Behera, A. D. Munshi, and A. Pal. 2010. Gynoecious inbred with better combining ability improves yield and earliness in bitter gourd (*Momordica charantia* L.). Euphytica 173:37–47

3. Dey, S. S., T. K. Behera, Anand Pal, and A. D. Munshi. 2005. Correlation and path coefficient analysis in bitter gourd (*Momordica charantia* L.). Veg Sci 32: 173-176.

4. Fan, Z., M. D. Robbins, and J. E. Staub. 2006. Population development by phenotypic selection with subsequent marker-assisted selection for line extraction in cucumber (*Cucumis sativus* L.). Theor Appl Genet 112:843–855

5. Robinson, R. W., and D. S. Decker-Walters. 1999. Cucurbits. CAB International, Wallingford, Oxford, UK. 240 pp.

Table 1. Per se performance of hybrids of bitter gourd for yield and its related traits

Hybrids	No of fruits/plant	Fruit length(cm)	Fruit diameter(cm)	Yield/plant (g)
DBGy- 201 x PDM	5.87	10.53	10.63	223.25
DBGy- 201 x PV	6.88	10.87	11.87	245.86
DBGy- 201 x Sel 2	5.38	9.60	10.40	197.92
DBGy- 201 x Nakhra Local	7.55	8.70	11.30	207.75
DBGy- 201 x MC84	7.12	10.60	10.80	240.21
DBGy- 201 x S26	6.87	9.70	10.40	176.74
DBGy- 201 x S29	5.88	7.67	11.13	245.45
DBGy- 201 x S30	3.62	7.13	9.57	235.77
DBGy- 201 x S41	6.96	7.50	9.90	216.33
DBGy- 201 x S54	12.79	11.07	10.57	216.98
DBGy- 201 x S57	7.82	9.07	10.60	226.70
DBGy- 201 x DBG 34	8.85	7.70	11.07	164.49
VNR 22 (Check)	5.42	10.77	12.20	172.26
US 33	4.45	9.80	7.20	144.73
PH 2	3.55	7.63	9.17	151.84
SE(d)	2.19	1.73	1.32	29.49
SE(m)	1.07	0.84	0.64	14.32
CD (P=0.05)	0.75	0.59	0.46	10.13
CV	19.76	11.16	7.53	8.58

The Distribution and Application of Bitter Gourd in China

Chun-mei Duan, Zhen Liu, Hong-wen Cui[1]
Zhongdu Seeds Science & Technology Co., Ltd (Shanghai, China)

Bitter gourd (*Momordica charantia* L.), also known as bitter melon, bitter cucumber or balsam pear. It is also sometimes called "elegant vegetable" because its bitterness does not affect other vegetables' flavor when cooked together. Bitter gourd belongs to the family of Cucurbitaceae, which is an annual climbing-vine. It originated from the tropical areas of Eastern India, and is now widely grows in tropical, subtropical and temperate zones, especially in India, Japan, Southeast Asia and China (1).

Bitter gourd with a cultivation history of more than 600 years is popular in China as a vegetable and as an ornament plant. As early as the Ming dynasty, the medicine book "Materia Medica for Famines" (1406) written by Zhu Su, mentioned bitter gourd, in another book "Nongzheng Quanshu" (1639) written by Guangqi Xu, mentioned that bitter gourd was very popular in southern China. These references indicate the bitter gourd grew widely in southern China during that period. Now this vegetable is grown in northern China as well, even as north as Shanxi, Shandong, Hebei, even Heilongjiang province.

The bitterness of bitter gourd derives from a special bitter glycoside. The glycoside is broken up when the fruit ripens (turns orange and mushy) so young bitter gourd is bitter, but the mature ones are sweet.

Bitter gourd is highly nutritious. It contains lots of vitamins, such as B1, B2, B3 and C. Also it has abundant magnesium, zinc, phosphorus, manganese, folic acid and dietary fiber. It has been reported that every 100 g fresh fruit contains 94 g water, 0.7 to 1.0 g protein, 2.6 to 3.5 g carbohydrates and 56 to 84 mg vitamins C (2). Bitter gourd is often used in Chinese cooking for its bitter flavor, typically in stir-fries (often with pork or eggs), soups, and teas. For example, in summer, bitter gourd is used for a bitter herbal tea in Guangzhou. In Nanjing, pieces of bitter gourd, rice and sugar are boiled together as a soup, also in the areas of Hubei and other places bitter gourd is used as a juice drink.

Bitter gourd is a kind of valuable medicine material. Its root, stem, leaf, flower, fruit and seeds all have medicinal uses. It has been used in traditional Chinese medicine for many years. According to the record of "Compendium of Materia Medica", written by Dr. Li Shizhen (1594), bitter gourd has a beneficial effect of removing the extra heat, eliminating fatigue, alleviating eye problems and improving eyesight. Some other traditional Chinese medicine scientists claimed that bitter gourd could stimulate digestion and relieve the problem of sluggish digestion, dyspepsia, and constipation. Recent research shows that bitter gourd can lower the level of blood glucose, rebuild the immune system and increase resistance against infection. Now in China, a variety of health care products containing bitter gourd has appeared in the market, such as bitter melon capsules, bitter glycosides and hypoglycemic agents (3).

Bitter gourd can be classified into several categories by size, shape and color. According to the size of the fruit; it can be divided into two groups: If the fruit is longer than 20 centimeters, it is called big bitter gourd, otherwise it is called small bitter gourd. According to fruit shape, it can be classified into three groups, long conical shape, short conical shape and long cylindrical shape. For example, Huashen bitter gourd has a long conical shape, and Dading bitter gourd has a short conical shape. Conical bitter gourd is mainly distributed in South China, such as Guangdong province. Fruit color can divided fruit into cyan and white. Cyan bitter gourd can be subdivided into dark green, green, jade green and celadon bitter gourd, while white bitter gourd can be subdivided into white and ivory bitter gourd.

Bitter gourd grows well in hot and humid climates, so it is mainly distributed in the Yangtze River basin and farther south. Guangdong, Guangxi, Fujian, Hainan, Jiangxi, Hunan and Sichuan provinces are the major cultivation areas of bitter gourd. Among different bitter gourd species, white bitter gourd is mainly distributed in the Yangtze River basin, while the cyan bitter gourd is mainly grown in Guangdong, Guangxi and Hainan provinces. Bitter gourd is becoming increasingly popular as a greenhouse crop and thus the production has gradually increased in North China where it is becoming a favorite vegetable (Fig. 1).

Farmers prefer varieties with early maturity, good disease resistance and high yields. Compared to other vegetables, the seed coat of bitter gourd is thicker. So, in order to improve germination rate, farmers often take some measures like soaking, breaking the shell, using chemicals, and so on. In the preliminary growth state, it is necessary to prune the extra branches and leave a main stem. However, above 1.5 meters, strong lateral branches can be left on the plant. Under normal circumstances, bitter gourd begins to flower at 40-45 days after sowing, and continue throughout a season of usually six months. Bitter gourd can be pollinated by insect like bees, or if grown in the greenhouse sometimes it needs hand pollination. After 15-20 days pollination, the fruit can be harvested. Picking too early or late will reduce the quality and yield of this crop.

Bitter gourd is a nutritious and healthful vegetable that is easy to produce because it seldom suffers damage from pests and diseases, thus it requires little pesticide application. These attributes are making this crop more and more popular in China.

Literature Cited

1. Grubben, G.J.H. 1977. Tropical vegetable and their genetic resources. IBPGR, Rome. pp. 51-52.

2. Jiang Xianming. 1984. Bitter gourd. In: Cultivation of vegetables. Press of China Agriculture. J. pp. 250-251.

3. Xie Chunsheng. 1994. The bitter gourd capsule's medical use for diabetes and the methods to prepare it. Patent 94111598.4 P, China.

Fig. 1. The distribution of the main varieties of bitter gourd in China.

Gene List 2010 for Cucumber

Adam D. Call and Todd C. Wehner

Department of Horticultural Science, North Carolina State University

This is the latest version of the gene list for cucumber (*Cucumis sativus* L.). In addition to morphological and resistance genes, this list includes genes that have been cloned from different plant tissues of cucumber. The genes in the list have been grouped into ten categories as follows: seedling markers, stem mutants, leaf mutants, flower mutants, fruit type mutants, fruit color mutants, resistance genes (mostly to diseases), protein (isozyme) variants, DNA (RFLPs and RAPDs) markers (Table 1), and cloned genes (Table 2). There is also a review of linkage of the morphological and resistance genes. As of 2010 the number of cloned genes of cucumber has become to vast for this article. Up to date information on cloned cucumber genes can be found at the National Center for Biotechnology Information Database (http://www.ncbi.nlm.nih.gov/). Complete lists and updates of genes for have been published previously, as follows:

Previous Gene Lists

- Robinson et al., 1976
- Robinson et al., 1982
- Pierce and Wehner, 1989
- Wehner, 1993
- Wehner and Staub, 1997
- Xie and Wehner, 2001
- Wehner, 2005

Revisions to the 2010 cucumber gene list include the addition of *Ch* for chilling resistance in seedlings (Kozik and Wehner, 2006, 2008) and *tf* for twin fused fruit (Klosinka et al., 2006). Researchers are encouraged to send reports of new genes, as well as seed samples to the cucumber gene curators (Yiqun Weng and Todd C. Wehner). Please inform us of omissions or errors in the gene list. Scientists should consult the list as well as the rules of gene nomenclature for the Cucurbitaceae (Robinson et al., 1976; Robinson et al., 1982) before choosing a gene name and symbol. That will avoid duplication of gene names and symbols. The rules of gene nomenclature were adopted in order to provide guidelines for naming and symbolizing genes. Scientists are urged to contact members of the gene list committee regarding rules and gene symbols.

Seed and seedling mutants

One of the advantages of using the cucumber in genetic research is the availability of seedling markers. To date, five non-lethal color mutants [virescent (*v*) (Poole, 1944; Tkachenko, 1935), variegated virescence (*vvi*) (Abul-Hayja and Williams, 1976), yellow cotyledons-1 (*yc-1*) (Aalders, 1959), yellow cotyledons-2 (*yc-2*) (Whelan and Chubey, 1973; Whelan et al., 1975), yellow plant (*yp*) (Abul-Hayja and Williams, 1976)] and 4 lethal, color mutants [chlorophyll deficient (*cd*) (Burnham et al., 1966), golden cotyledon (*gc*) (Whelan, 1971), light sensitive (*ls*) (Whelan, 1972b), pale lethal (*pl*) (Whelan, 1973)] have been identified.

Six seedling traits which affect traits other than color include bitterfree (*bi*) (Andeweg and DeBruyn, 1959), blind, (*bl*) (Carlsson, 1961), delayed growth (*dl*) (Miller and George, 1979), long hypocotyl (*lh*) (Robinson and Shail., 1981), revolute cotyledons (*rc*) (Whelan et al., 1975) and stunted cotyledons (*sc*) (Shanmugasundarum and Williams, 1971; Shanmugasundarum et al., 1972).

Seedling chilling resistance is conferred by the dominant gene *Ch* (Kozik and Wehner, 2006, 2008).

Stem mutants

Seven genes have been identified which affect stem length: bush (*bu*) (Pyzenkov and Kosareva, 1981), compact (*cp*) (Kauffman and Lower, 1976), determinate (*de*) (Denna, 1971; George, 1970; Hutchins, 1940), dwarf (*dw*) (Robinson and Mishanec, 1965), tall height (*T*) (Hutchins, 1940) and *In-de* which behaves as an intensifier for *de* (George, 1970). Rosette (*ro*) which also affects height is characterized by muskmelon-like leaves (de Ruiter et al., 1980).

Unlike these genes, fasciated (*fa*) (Robinson, 1978b; Shifriss, 1950) affects stem confirmation, not length.

Leaf mutants

Several genes have been shown to control leaf or foliage characteristics. Eight in particular are responsible for leaf shape: blunt leaf apex (*bla*) (Robinson, 1987a), cordate leaves-1 (*cor-1*) (Gornitskaya, 1967), cordate leaves-2 (*cor-2*) (Robinson, 1987c), crinkled leaf (*cr*) (Odland and Groff, 1963a), divided leaf (*dvl*) (den Nijs and Mackiewicz, 1980), ginko leaf (*gi*) (John and Wilson, 1952), little leaf (*ll*), (Goode et al., 1980; Wehner et al., 1987) and umbrella leaf (*ul*) (den Nijs and de Ponti 1983). Note that ginko leaf is a misspelling of the genus *Ginkgo*.

The original cordate leaf gene identified by Gornitskaya (1967) differs from *cor* proposed by (Robinson, 1987c) which also had calyx segments which tightly clasp the corolla, hindering flower opening and insect pollination. Therefore, we propose that the first gene identified by Gornitskaya be labeled *cor-1* and the second identified by Robinson be labeled *cor-2*. It should be noted that plants with stunted cotyledon may look similar to those with ginko at the younger stages but the cotyledons of *sc* mutants are irregular and *gi* mutants are sterile.

Opposite leaf arrangement (*opp*) is inherited as a single recessive gene with linkages to *m* and *l*. Unfortunately, incomplete penetrance makes the opposite leaf arrangement difficult to distinguish from normal plants with alternate leaf arrangement (Robinson, 1987e).

Five mutants which affect color or anatomical features of the foliage are golden leaves (*g*) (Tkachenko, 1935), glabrous (*gl*) (Inggamer and de Ponti, 1980; Robinson and Mishanec, 1964), glabrate (*glb*) (Whelan, 1973), short petiole (*sp*) (den Nijs and Boukema, 1985) and tendrilless (*td*) (Rowe and Bowers, 1965).

Flower mutants

Sex expression in cucumber is affected by several single-gene mutants. The *F* locus affects gynoecy (femaleness), but is modified by other genes and the environment, and interacts with *a* and *m* (androecious and andromonoecious, respectively) (Galun, 1961; Kubicki, 1969; Rosa, 1928; Shifriss, 1961; Tkachenko, 1935; Wall, 1967). Androecious plants are produced if *aa* and *ff* occur in combination, otherwise plants are hermaphroditic if *mm FF*, andromonoecious if *mm ff*, gynoecious if *MM FF* and monoecious if *MM ff*. The gene *F* may also be modified by an intensifier gene *In-F* which increases the femaleness (Kubicki, 1969b). Other genes that affect sex expression are *gy* for gynoecious (Kubicki, 1974), *m-2* for andromonoecious (Kubicki, 1974) and *Tr* for trimonoecious expression (Kubicki, 1969d).

Cucumbers, typically considered day-neutral plants, have occasionally been shown to express sensitivity to long days. Della Vecchia et al. (1982) and Shifriss and George (1965) demonstrated that a single gene for delayed flowering (*df*) is responsible for this short-day response.

Another gene which may give the impression of eliciting daylength sensitivity by causing a delay in flowering is *Fba*. In reality, *Fba* triggers flower bud abortion

prior to anthesis in 10 to 100% of the buds (Miller and Quisenberry, 1978).

Three separate groups have reported single genes for multiple pistillate flowers per node. Nandgaonkar and Baker (1981) found that a single recessive gene *mp* was responsible for multiple pistillate flowering. This may be the same gene which Fujieda et al. (1982) later labeled as *pf* for plural pistillate flowering. However, they indicated that 3 different alleles were responsible, with single pistillate being incompletely dominant over multiple pistillate: pf^+ for single pistillate, pf^d for double pistillate and pf^m for multiple pistillate (more than 2 flowers per node).

Thaxton (1974), reported that clustering of pistillate flowers is conditioned by a single dominant gene (we propose the symbol, *Mp-2*), and that modifier genes influence the amount of clustering. Thaxton (1974) also determined that clustering of perfect flowers is controlled by genes different from clustering of gynoecious flowers.

Several genes for male sterility have been reported for cucumber, but because of the ease of changing sex expression with growth regulators, little commercial use has been made of them. Five genes, *ms-1*, *ms-2*, *ap*, *cl* and *gi* have been identified. The genes *ms-1* and *ms-2* cause sterility by pollen abortion before anthesis; *ms-1* plants are also partially female sterile (Robinson and Mishanec, 1965; Shanmugasundarum and Williams, 1971; Whelan, 1972a). Apetalous mutants (*ap*) on the other hand have infertile anthers which appear to have been transformed into sepal-like structures (Grimbly, 1980). Ginko (*gi*), mentioned earlier as a leaf mutant, also causes male sterility (John and Wilson, 1952).

One of these male steriles may be of little use except as a genetic marker. Closed flower (*cl*) mutants are both male and female sterile, so seed production must be through the heterozygotes only (Groff and Odland, 1963). With this mutant, the pollen is inaccessible to bees because the buds remain closed.

Three genes alter floral characteristics: green corolla (*co*) (Currence, 1954; Hutchins, 1935), orange-yellow corolla (*O*) (Tkachenko 1935), negative geotropic peduncle response (*n*) (Odland and Groff, 1963b). Green corolla (*co*), named because of its green petals, has enlarged but sterile pistils (Currence, 1954; Hutchins, 1935), and has potential for use as a female sterile in hybrid production.

Fruit mutants

Because the fruit is the most important part of the cucumber economically, considerable attention has been given to genes affecting it. One such gene is Bitter fruit,

Bt, (Barham, 1953) which alters fruit flavor by controlling cucurbitacin levels. The gene *Bt* is different from *bi* because it consistently alters only the fruit cucurbitacin levels compared to *bi* which affects the whole plant.

Five genes conditioning skin texture are *Tu* (Strong, 1931; Wellington, 1913), *te* (Poole, 1944; Strong, 1931), *P* (Tkachenko, 1935), *I* (Tkachenko, 1935) and *H* (Hutchins, 1940; Tkachenko, 1935). Smooth (*Tu*) and tender (*te*) skin are usually associated with European types, while American types are generally warty and thick skinned (Poole, 1944; Strong 1931). Heavy netting, *H*, which occurs when fruit reach maturity may be tightly linked or pleiotropic with *R* and *B* (discussed later).

In *Cucumis sativus* var. *tuberculatus*, Tkachenko (1935) found that gene *P*, causing fruit with yellow rind and tubercles, was modified by gene *I*, an intensifier which increases the prominence of the tubercles (Tkachenko, 1935).

There are 3 genes which affect internal fruit quality, each identified by viewing transections of fruits; Empty chambers-1 (*Es-1*), Empty chambers-2 (*Es-2*) (Kubicki and Korzeniewska, 1983) and locule number (*l*) (Youngner, 1952).

Hutchins (1940) proposed that 2 genes controlled spine characteristics, with *f* producing many spines and being tightly linked with *s* which produced small spines. Poole (1944) used the data of Hutchins (1940) to suggest that *s* and *f* were the same gene and proposed the joint symbol *s* for a high density of small spines. Tkachenko (1935) who used the same symbol for control of less dense spines, did not look at spine size, and the same gene might have been involved. However, Fanourakis (1984) and Fanourakis and Simon (1987) reported 2 separate genes involved, and named them *ss* and *ns* for small spines and numerous spines, respectively.

These may differ from those that led Carruth (1975) to conclude that 2 genes act in a double recessive epistatic fashion to produce the dense, small spine habit. We propose that these genes be labeled *s-2* and *s-3* and *s1* be used instead of *s* proposed by Poole (1944).

Carruth (1975) and Pike and Carruth (1977) also suggested that carpel rupture along the sutures was inherited as a single recessive gene that was tightly linked with round, fine-spined fruits. This may be similar to what Tkachenko (1935) noted in the 'Klin mutant' as occasional deep-splitting flesh. We suggest the symbol *cs* for carpel splitting, but note that because penetrance of the trait may be lower under certain environmental conditions (Carruth, 1975) this trait may be related to the gooseberry (*gb*) fruit reported by Tkachenko (1935). Another character not found in commercial cultivars was protruding ovary (*pr*) reported by Youngner (1952).

There is dispute over the inheritance of parthenocarpy, a trait found in many European cucumbers (Wellington and Hawthorn, 1928). Pike and Peterson (1969) suggested an incompletely dominant gene, *Pc*, affected by numerous modifiers, was responsible. In contrast, de Ponti and Garretsen (1976) explained the inheritance by 3 major isomeric genes with additive action.

A modifier of fruit length, *Fl*, was identified by its linkage with scab resistance (*Cca*) (Henry Munger, personal communication; Wilson, 1968). Expressed in an additive fashion, fruit length decreases incrementally from heterozygote to homozygote (*fl fl*).

A gene for inheritance of twin fused fruit, *tf*, was discovered in gynoecicous inbred B 5263 (Klosinska et al., 2006). The trait is characterized by pairs of two separate pistallate flowers with partially joined ovaries on a single peduncle at a node developing into twin fused fruit during development. Twin fused fruit were only observed on gynoecious plants, indicating epistatis.

Fruit Color

Twelve mutants have been identified which affect fruit color either in the spines, skin, or flesh and a few of these appear to act pleiotropically. For example, *R* for red mature fruit color is very closely linked or pleiotropic to *B* for black or brown spines and *H* for heavy netting (Hutchins, 1935; Tkachenko, 1935; Wellington, 1913). It also interacts with *c* for cream colored mature fruit in such a way that plants which are (*RR CC*), (*RR cc*), (*rr CC*) and (*rr cc*) have red, orange, yellow and cream colored fruits, respectively (Hutchins, 1940).

The *B* gene produces black or brown spines and is pleiotropic to or linked with *R* and *H* (Wellington, 1913). The homozygous recessive plant is white spined with cream colored mature fruit and lacks netting. Other spine color genes are *B-2*, *B-3* and *B-4* (Cowen and Helsel, 1983; Shanmugasundarum et al., 1971a).

White immature skin color (*w*) is recessive to the normal green (Cochran, 1938), and yellow green (*yg*) is recessive to dark green and epistatic with light green (Youngner, 1952). Skin color may also be dull or glossy (*D*) (Strong, 1931; Tkachenko, 1935) and uniform or mottled (*u*) (Andeweg, 1956; Strong, 1931).

Kooistra (1971) reported 2 genes that affect fruit mesocarp color. White flesh (*wf*) and yellow flesh (*yf*) gene loci interact to produce either white (*WfWf YfYf* or *wfwf YfYf*), yellow (*WfWf yfyf*), or orange (*wfwf yfyf*) flesh color.

Insect Resistance

Bitterfree, *bi* (Andeweg and DeBruyn, 1959), is responsible for resistance to spotted and banded cucumber beetles (*Diabrotica* spp.) (Chambliss, 1978; Da Costa

& Jones, 1971a; Da Costa & Jones, 1971b) and two-spotted spider mites (*Tetranychus urticae* Koch.) (Da Costa & Jones, 1971a; Soans et al., 1973). However, this gene works inversely for the 2 species. The dominant allele which conditions higher foliage cucurbitacin levels incites resistance to spider mites by an antibiotic affect of the cucurbitacin. The homozygous recessive results in resistance to cucumber beetles because cucurbitacins are attractants.

In the 1989 Cucurbit Genetics Cooperative Report the authors labeled the gene for resistance to *Diabrotica* spp. *di*, but wish to retract it in light of recent evidence.

Disease Resistance

Currently there are 15 genes known to control disease resistance in *C. sativus*. Three of these condition virus resistance. Wasuwat and Walker (1961) found a single dominant gene, *Cmv*, for resistance to cucumber mosaic virus. However, others have reported more complex inheritance (Shifriss et al., 1942). Two genes condition resistance to watermelon mosaic virus, *Wmv* (Cohen et al, 1971) and *wmv-1-1* (Wang et al., 1984). Most recently, resistance to zucchini yellow mosaic virus (*zymv*) has been identified (Provvidenti, 1985).

Both resistance to scab, caused by *Cladosporium cucumerinum* Ell. & Arth., and resistance to bacterial wilt caused by *Erwinia tracheiphila* (E. F. Smith) Holland are dominant and controlled by *Ccu* (Abul-Hayja et al., 1978; Andeweg, 1956; Bailey and Burgess, 1934) and *Bw* (Nuttall and Jasmin, 1958; Robinson and Whitaker, 1974), respectively. Other dominant genes providing resistance are: *Cca* for resistance to target leaf spot (*Corynespora cassiicola*) (Abul-Hayja et al., 1978), *Cm* for resistance to Corynespora blight (*Corynespora melonis*) (van Es, 1958), *Foc* for resistance to Fusarium wilt (*Fusarium oxysporum* f. sp. *cucumerinum*) (Netzer et al., 1977) and *Ar* for resistance to anthracnose [*Colletotrichum lagenarium* (Pars.) Ellis & Halst.] (Barnes and Epps, 1952). In contrast, resistance to *Colletotrichum lagenarium* race 1 (Abul-Hayja et al., 1978) and angular leaf spot (*Pseudomonas lachrymans*) (Dessert et al., 1982) are conditioned by the recessive genes *cla* and *psl*, respectively.

Several reports have indicated that more than one gene controls resistance to powdery mildew [*Sphaerotheca fuliginea* (Schlecht) Poll.] with interactions occurring among loci (Hujieda and Akiya, 1962; Kooistra, 1968; Shanmugasundarum et al., 1971b). The resistance genes *pm-1* and *pm-2* were first reported by Hujieda and Akiya (1962) in a cultivar which they developed and named 'Natsufushinari'. Kooistra (1968) using this same cultivar, later confirmed their findings and identified one additional gene (*pm-3*) from USDA accessions PI200815

and PI200818. Shimizu et al. (1963) also supported 3 recessive genes which are responsible for resistance of 'Aojihai' over 'Sagamihan'.

Several genes with specific effects have been identified more recently (Shanmugasundarum et al., 1971b) but unfortunately, direct comparisons were not made to see if the genes were identical with *pm-1*, *pm-2* and *pm-3*. Fanourakis (1984) considered a powdery mildew resistance gene in an extensive linkage study and proposed that it was the same gene used by Shanmugasundarum et al. (1971b) which also produces resistance on the seedling hypocotyl. Because expression is identified easily and since it is frequently labeled in the literature as '*pm*' we believe that this gene should be added to the list as *pm-h* with the understanding that this may be the same as *pm-1*, *pm-2* or *pm-3*.

There are several proposed inheritance patterns for resistance to downy mildew. They range from three recessive genes (Doruchowski and Lakowska-Ryk, 1992; Shimizu et al., 1963) to three partially dominant genes (Pershin et al., 1988) to an interaction between dominant susceptible and recessive resistance genes (Badr and Mohamed, 1998; El-Hafaz et al., 1990) to one or two incompletely dominant genes (Petrov et al., 2000) to a single recessive gene (Angelov, 1994; Fanourakis and Simon, 1987; Van Vliet and Meysing, 1974; 1976). Doruchowski and Lakowska-Ryk (1992) had evidence that downy mildew resistance was controlled by three recessive genes (*dm-1*, *dm-2* and *dm-3*), where *dm-3* and either *dm-1* or *dm-2* had to be homozygous recessive for maximum resistance. However, there was discrepancy in the F_2 results, which did not agree with their model. They argued that this resulted from testing too narrow a population. The three genes were included in previous cucumber gene lists but have been removed from the current list as none of the genes were identified and no type lines are available to use in studies of separate genes. A review of the inheritance of downy mildew resistance is in Criswell (2008).

One gene, *dm* (*dm-1*), has been identified which confers resistance to downy mildew [*Pseudoperonospora cubensis* (Berk. & Curt.) Rostow] (van Vliet and Meysing, 1974). Inherited as a single recessive gene, it also appeared to be linked with *pm* (van Vliet, 1977). The *dm* gene traces its resistance to PI 197087, a wild accession collected in India. There are, however, indications that more than one gene may be involved (Jenkins, 1946). Angelov (1994) reported that PI 197088 resistance was due to two recessive genes. PI 197088 was collected from the same region and at the same time as PI 197087. PI 197088 was recently reported as highly resistant to the current downy mildew in the southeastern U.S., while cultigens tracing resistance to PI 197087 are no longer highly resistant (Call, 2010). Interestingly, it appears the 2004 change of the downy mildew population resulted in a change in rank of resistant and moderately resistant cultigens. Previously highly resistant cultignes are now moderate, while cultigenes that are now highly resistant were only moderate prior to 2004 (Call and Wehner, 2010). The first reported source of resistance was from a Chinese cultigen tested at the Puerto Rico Agricultural Experiment Station in 1933. This cultigen was crossed with the best available commercial cultivars to combine the resistance with good horticultural traits. This eventually led to seven resistant lines having good characteristics. Of these, PR 37, PR 39, and PR 40 had high quality fruit and yield superior to commercial checks. These were used in further development of resistant breeding lines and cultivars (Barnes et. al. 1946; Barnes, 1955; Barnes and Epps, 1955). It appears that there are at least three genes for resistance to downy mildew in cucumber: one from the Chinese cultivar used in developing the PR lines, one from PI 197087, and one from PI 197088 (assuming that PI 197087 and PI 197088 share one resistance gene, *dm-1*).

Multiple genes for virus resistance have been reported, including genes conferring resistance to CMV, MWM, PRSV, WMV, and ZYMV. The resistance gene *cmv* was identified in 'National Pickling' and 'Wis. SR 6' for resistance cucumber mosaic virus was reported by Wasuwat and Walker (1961). Kabelka and Grumet (1997) reported Moroccan watermelon mosaic virus resistance as a single recessive gene, *mwm*, from Chinese cucumber cultivar 'TMG-1'. Resistance to papaya ringspot virus (formerly watermelon mosaic virus 1) was identified in the cultivar 'Surinam' by Wang et al. (1984). Multiple genes for resistance to watermelon mosaic virus (WMV) have been reported. Cohen et al. (1971) identified a dominant gene, *Wmv*, in the cultivar 'Kyoto 3 Feet', resistant to strain 2 of WMV. The gene *wmv-1-1* from the cultivar 'Surinam' was reported by Wang et al. (1984) to give resistance to strain 1 of WMV by means of limiting systemic translocation. Cultigens having wmv-1-1 may show severe symptoms on lower leaves, while symptoms on the rest of the plant are limited. Wai et al. (1997) reported three genes for watermelon mosaic virus resistance from the cultigen TMG-1: *wmv-2*, *wmv-3*, and *wmv-4*. Expression of *wmv-3* and *wmv-4* is limited to true leaves only, while *wmv-2* is expressed throughout the plant. Two genes have been reported conferring resistance to zucchini yellow mosaic virus: *zym-Dina* from the cultigen Dina-1 and zym-TMG1 from the cultigen TMG-1. Inheritance of *zym-TMG1* is incomplete but typically recessive.

Environmental Stress Resistance

Two genes have been identified for stress resistance, resistance to sulfur dioxide air pollution conditioned by *Sd* (Bressan et al., 1981) and increased tolerance to high salt levels conditioned by the gene, *sa*, Jones (1984).

Other Traits

The dominant allele, *Psm*, induces paternal sorting of mitochondria, where *Psm* is from MSC 16 and *psm* is from PI 401734 (Havey et al., 2004).

Molecular and Protein Markers

Isozyme variant nomenclature for this gene list follows the form according to Staub et al. (Staub et al., 1985), such that loci coding for enzymes (e.g. glutamine dehydrogenase, G2DH) are designated as abbreviations, where the first letter is capitalized (e.g. G2dh). If an enzyme system is conditioned by multiple loci, then those are designated by hyphenated numbers, which are numbered from most cathodal to most anodal and enclosed in parentheses. The most common allele of any particular isozyme is designated 100, and all other alleles for that enzyme are assigned a value based on their mobility relative to that allele. For example, an allele at locus 1 of FDP (fructose diphosphatase) which has a mobility 4 mm less that of the most common allele would be assigned the designation *Fdp(1)-96*.

RFLP marker loci were identified as a result of digestion of cucumber DNA with *Dra*I, *Eco*RI, *Eco*RV, or *Hind*III (Kennard et al., 1994). Partial-genomic libraries were constructed using either *Pst*I-digested DNA from the cultivar Sable and from *Eco*RV-digested DNA from the inbred WI 2757. Derived clones were hybridized to genomic DNA and banding patterns were described for mapped and unlinked loci (CsC482/H3, CsP314/E1, and CsP344/E1, CsC477/H3, CsP300/E1).

Clones are designated herein as CsC = cDNA, CsP = *Pst*I-genomic, and CsE = *Eco*RI-genomic. Lower-case a or b represent two independently-segregating loci detected with one probe. Lower-case s denotes the slowest fragment digested out of the vector. Restriction enzymes designated as DI, *Dra*I; EI, *Eco*RI; E5, *Eco*RV; and H3, *Hind*III. Thus, a probe identified as CsC336b/E5 is derived from a cDNA library (from 'Sable') which was restricted using the enzyme *Eco*RV to produce a clone designated as 336 which displayed two independently segregating loci one of which is b. Clones are available in limited supply from Jack E. Staub.

RAPD marker loci were identified using primer

sequences from Operon Technologies (OP; Alameda, California, U.S.A.) and the University of British Columbia (Vancouver, BC, Canada). Loci are identified by sequence origin (OP or BC), primer group letter (e.g., A), primer group array number (1-20), and locus (a, b, c, etc.) (Kennard et al., 1994). Information regarding unlinked loci can be obtained from Jack E. Staub.

Because of their abundance, common source (two mapping populations), and the accessibility of published information on their development (Kennard et al., 1994) DNA marker loci are not included in Table 1, but are listed below.

The 60 RFLP marker loci from mapping cross Gy 14 x PI 183967 (Kennard et al., 1994): CsP129/E1, CsC032a/E1, CsP064/E1, CsP357/H3, CsC386/E1, CsC365/E1, CsP046/E1, CsP347/H3, CsC694/E5, CsC588/H3, CsC230/E1, CsC593/D1, CsP193/H3, CsP078s/H3, CsC581/E5, CsE084/E1, CsC341/H3, CsP024/E1, CsP287/H3, CsC629/H3, CsP225s/E1, CsP303/H3, CsE051/H3, CsC366a/E5, CsC032b/E1, CsP056/H3, CsC378/E1, CsP406/E1, CsP460/E1, CsE060/E1, CsE103/E1, CsP019/E1, CsP168/D1, CsC560/H3, CsP005/E1, CsP440s/E1, CsP221/H3, CsC625/E1, CsP475s/E1, CsP211/E1, CsP215/H3, CsC613/E1, CsC029/H3, CsP130/E1, CsC443/H3, CsE120/H3, CsE031/H3, CsC366b/E5, CsC082/H13, CsP094/H3, CsC362/E1, CsP441/E1, CsP280/H3, CsC137/H3, CsC558/H3, CsP037a/E1, CsP476/H3, CsP308/E1, CsP105/E1, and Csc166/E1.

The 31 RFLP marker loci from mapping cross Gy 14 x PI 432860 (Kennard et al., 1994): CsC560/D1, CsP024/E5, CsP287/H3, CsC384/E5, CsC366/E5, CsC611/D1, CsP055/D1, CsC482/H3, CsP019/E1, CsP059/D1, CsP471s/H13, CsC332/E5, CsP056/H3, CsC308/E5, CsP073/E5, CsP215/H3, CsC613/D1, CsP266/D1, CsC443/H3, CsE031/E1, CsE120/H3, CsE063/E1, CsP444/E1, CsC612/D1, Cs362/E1, CsP280/H3, CsC558/H3, CsP008/D1, CsP308/E1, CsC166/E1, and CsP303/H3.

The 20 RAPD marker loci from mapping cross Gy 14 x PI 432860 (Kennard et al., 1994): OPR04, OPW16, OPS17, OPE13a, OPN06, OPN12, OPP18b, BC211b, OPN04, OPA10, OPE09, OPT18, OPA14b, OPU20, BC460a, OPAB06, OPAB05, OPH12, OPA14a, and BC211a.

In addition to the isozymes, RFLPs and RAPDs, nearly 100 cloned genes are listed here (Table 2).

Possible Allelic or Identical Genes

Several of the genes listed may be either pleiotropic, closely linked, or allelic. Additional research is needed to compare the sources of the various similar genes to ensure that they are not duplicates. In some instances this may be difficult because many of the earlier publications did not list the source of the genes or the methods used to measure the traits, and many of these authors are deceased.

An example of this is the two-locus model (R c) for fruit color. We have been unable to locate any plants with red or yellow colored mature fruits. All plants evaluated in other studies have color inherited as a single gene. Hutchins may have separated fruit with cream color into 2 groups, yellow and cream, and fruits with orange color into two groups, orange and red. However, those distinctions are difficult to make using available germplasm. Situations like these may be impossible to resolve. Red mature fruit may be the same as the brownish-red cracked or netted exocarp found in some accessions such as PI 330628. Orange fruit occur in black spined accessions such as Wis. SMR 18.

In the future, researchers should use the marker lines listed here, or describe and release the marker lines used so that allelism can be checked by others. Currently, groups of similar genes that need to be checked to determine how they are related include the following: the chlorophyll deficiency mutants (cd, g, ls, pl, v, vvi, yc-1, yc-2, and yp), the stem mutants (bu, de, dw, In-de, and T), the leaf shape mutants (rc and ul), the sex expression mutants (a, F, gy, In-F, m, m-2, and Tr), the male sterility genes (ap, cl, ms-1, and ms-2), the flowering stage mutants (df and Fba), the flower color mutants (co and O), the powdery mildew resistance mutants (pm-1, pm-2, pm-3 and pm-h), the fruit spine color mutants (B, B-2, B-3, and B-4), the fruit skin color mutants (c, R, and w), the spine size and density mutants (s, s2, and s-3) and the seed cell mutants (cs and gb).

Two groups of associated traits, one from 'Lemon' cucumber (m, pr, and s) and the other involving fruit skin color, surface texture, and spine type (R, H, and B), need to be checked using large populations to determine whether they are linked or pleiotropic. Recent gains have been made in this area by Robinson (1978a) who demonstrated that the m gene is pleiotropic for fruit shape and flower type, producing both perfect flowers and round fruits, and Abul-Hayja et al. (1975) and Whelan (1973) who determined that gl and glb are independent genes.

New information indicates that comparisons also need to be made between resistance to scab (Ccu) and Fusarium wilt (Foc) and between resistance to target leaf spot (Cca) and Ulocladium cucurbitae leafspot. Mary Palmer (personal communication) found a fairly consistent association between resistances to scab and Fusarium wilt, which suggests that they might be linked or using the same mechanism for defense against the pathogen.

Similar defense mechanisms might also be responsible for similarities in resistance to target leaf spot (*Cca*) and *Ulocladium cucurbitae* leafspot (Henry Munger, personal communication).

Genetic Linkage

Since cucumber has just 7 chromosome pairs and over 100 known genes, it would seem that linkage maps would be fairly complete by now. Unfortunately, we know of few references reporting linkages of more than 2 gene loci, and this is the first review to summarize the literature for linkages and attempt to describe different linkage groups.

Many difficulties were encountered and should be considered when reading this review. First, a portion of the nomenclature is still unclear and some of the genes may be duplicates of others since common parents were not compared. This problem was discussed in the previous section. Secondly, some of the linkage relationships analyzed in previous studies did not involve specific genes. Linkages in several reports were discussed for plant traits that might have been inherited in multigenic fashion, or if a single gene were involved, it was not specifically identified.

Therefore, in this review linkages for traits without genes will be omitted and a '?' will follow each gene which has a questionable origin. Six linkage groups could be determined from the current literature (Fig. 1). The order in which the genes were expressed in each group does not necessarily represent the order in which they may be found on the chromosome.

Linkage Group A

The largest linkage group in cucumber has 14 genes, composed of *wmv-1-1, wmv-2, Prsv-2, gy, gl, dl, dvl, de, F, ms-2, glb, bi, df* and *B-3* or *B-4*. In contributing to this grouping, Whelan (1974) noted that *ms-2* is linked with *glb* (rf=.215±.029) and *de* (rf=.335±.042) while being independent of *bi, gl, yc-1, yc-2*, and *cr*. Gene *de* is linked with *F* (Odland and Groff, 1963b; Owens and Peterson, 1982) which in turn is linked with *B-3* or *B-4* (Cowen and Helsel, 1983), *gy* (rf=.04) (Kubicki, 1974), *bi* (rf=.375) and *df* (rf=34.7) (Fanourakis, 1984; Fanourakis and Simon, 1987). Gene *de* is also weakly linked with *dl* (Miller and George, 1979), strongly linked with *dvl* (Netherlands, 1982), and independent of *cp* (Kauffman and Lower, 1976). Gene *wmv-1-1* is linked with bitterfree (*bi*) but independent of *Ccu, B, F* or *pm* (Wang et al., 1987). Wai et al. (1997) showed linkage between *Prsv-2* and *bi* (rf=.28), *bi* and *F* (rf=.34), and *F* and *wmv-2* (rf=.33).

Two reports show that *dvl* is weakly linked with *gl* (rf=.40) and independent of *bi* and *Ccu* (Netherlands, 1982; den Nijs and Boukema, 1983), while Robinson

(1978f) originally indicated that *gl* was linked with *yc* and independent of *B, m, l*, and *yg* as well as *bi* (Netherlands, 1982) and *sp* (den Nijs and Boukema, 1985), but more recently he indicated that *gl* was independent of *yc* (Robinson, 1987d).

Completing linkage group A, Cowen and Helsel (1983) demonstrated that the spine color genes (*B-3* and *B-4*) were independent of the genes for bitterness, and Whelan (1973) found that *pl* was independent of *glb* and *bi*, while *glb* was independent of *gl, bi, ls, yc*, and *cr*. The last clarifies that *gl* and *glb* must indeed be separate loci.

Linkage Group B

Group B is composed of 9 genes (*n, pr, l, m, opp, m-2, Bw, s*? and *ms*?) unless *s*? (Robinson, 1978) is the same as *s* from Hutchins (1940) and Poole (1944). If these are the same, then linkage groups II and III will be joined for a total of 12 genes. Of the first 7, two pairs have been defined with recombination values. Youngner (1952) determined that *m* and *l* were linked with a recombination frequency of .326 ± .014 and Robinson determined that *opp* was linked to both (Robinson, 1987e). Iezzoni and Peterson (1979, 1980) found that *m* and *Bw* were separated by only one map unit (rf=.011±.003). Iezzoni et al. (1982) also determined that *m-2* was closely linked with both *m* and *Bw*, and that *Bw* was independent of *F* from linkage group I (Iezzoni and Peterson, 1980).

Robinson (1978c, 1978d), and Youngner (1952) found that linkages existed between *m, l, n, pr* and spine number (*s*?) with the possibility of pleiotropy being responsible for the *m/pr* relationship. They also demonstrated that *B, yg*, and *pm*? were independent of the same genes (Robinson, 1978c; Youngner, 1952).

Rounding out the linkage group is one of the male sterility genes (*ms*?). Robinson (1978d) found that it was linked with both *m* and *l*, but did not identify which male sterile gene it was.

Linkage Group C

Group C is the oldest and most mystifying linkage group. It is currently composed of *R* for red or orange mature fruit color, *H* for heavy netting, *B* for black or brown spine color, *c* for cream mature fruit color and *s* for spine frequency and size (Hutchins, 1940; Poole, 1944; Strong, 1931; Tkachenko, 1935). However, there is speculation on the nature of this linkage group. Since very few recombinants of the *R, H, B* and *c, h, b* linkage groups have been reported, it is also felt that these characteristics may be the response of 2 alleles at a single pleiotropic gene. There is also speculation that *R* and *c* are different alleles located at the same locus (see earlier discussion).

Hutchins (1940) found that *s* was independent of *B* and *H* while *s* was linked with *R* and *c*. If he was correct, then pleiotropy of *H* and *B* with *R* and *c* is ruled out. His report also indicated that *B* and *s* were independent of *de* as was *de* of *R*, *c* and *H*.

A possibility exists that this linkage group may be a continuation of group II through the *s* gene. Poole (1944) used the data of Hutchins (1940) to determine that *c* and *s* are linked with a recombination frequency of .163 ± .065. The question that remains is whether *s* (Hutchins, 1940; Poole, 1944) is the same as the gene for spine number in the findings of Robinson (1978c). If Cowen and Helsel (1983) are correct in their finding that a linkage exists between *F* and *B* then groups I and III may be on the same chromosome. However, in this text they will remain separated based on conclusions of Fanourakis (1984) which indicate that errors may be common when attempting to distinguish linkages with *F* since classification of *F* is difficult. This difficulty may also explain many conflicting reports.

Linkage Group D

Twelve genes (*ns, ss, Tu, Pc, D, U, te, cp, dm, Ar, coca* and *pm*? or *pm-h*) are in group D, but the identity of the specific gene for powdery mildew resistance is elusive. Van Vliet and Meysing (1947, 1977) demonstrated that the gene for resistance to downy mildew (*dm*) was either linked or identical with a gene for resistance to powdery mildew (*pm*?), but because the linkage between *pm*? and *D* was broken while that of *dm* and *D* was not, *pm*? and *dm* must be separate genes. The problem lies in the lack of identity of *pm*? because Kooistra (1971) also found that a gene for powdery mildew resistance (*pm*?) was linked to *D*.

Further complicating the identity of *pm*, Fanourakis (1984) found that *pm-h* was linked to *te* and *dm*, yet *cp*, which must be located at approximately the same locus, was independent of *te*. He suggested that there were either 2 linkage groups, *ns, ss, Tu, Pc, D, U, te* and *cp, dm, Ar*, located at distal ends of the same chromosome with *pm-h* at the center, or the 2 groups are located on different chromosomes with a translocation being responsible for apparent cross linkages. However, evidence for the latter which suggested that *F* was associated with the 7-gene segment is not probable since there are few other supportive linkages between genes of this segment and linkage group I. A more likely explanation is the occurrence of 2 or more genes conditioning resistance to powdery mildew being found on this chromosome.

More recently Lane and Munger (1985) and Munger and Lane (1987) determined that a gene for resistance to powdery mildew (*pm*?) was also linked with *coca* for susceptibility to target leaf spot but that linkage, though fairly tight, was breakable.

The last 4 genes in this group are *Tu, D, te* and *u* (Strong, 1931). Until recently it was believed that each in the recessive form were pleiotropic and consistent with European type cucumbers and each in the dominant form were pleiotropic and consistent with American type cucumbers. Fanourakis (1984) and Fanourakis and Simon (1987) reported that crossing over (R=23.7) occurred between *te* and the other 3 genes which still appeared to be associated. However, using triple backcrosses they demonstrated that there is a definite order for *Tu, D* and *u* within their chromosome segment and that the *Tu* end is associated with the *ns* and *ss* end.

Linkage Group E

Group E is currently composed of 3 genes *lh, sp* and *ul*. The gene *sp* was strongly linked with *lh* and weakly linked with *ul* (Zijlstra and den Nijs, 1986). However Zijlstra and den Nijs (1986) expressed concern for the accuracy of the *sp* and *ul* linkage data since it was difficult to distinguish *ul* under their growing conditions.

Linkage Group F

Group F is comprised of 2 genes, *Fl* and *Ccu* which appear to be tightly associated. Wilson (1968) concluded that pleiotropy existed between scab resistance and fruit length because backcrossing scab resistance into commercial varieties consistently resulted in reduced fruit length. However, Munger and Wilkinson (1975) were able to break this linkage producing varieties with scab resistance and longer fruit (Tablegreen 65 and 66, Marketmore 70 and Poinsett 76). Now when these varieties are used to introduce scab resistance long fruit length is consistently associated.

Unaffiliated Genes

Independent assortment data are as important in developing linkage maps as direct linkage data and several researchers have made additional contributions in this area. One of the most extensive studies, based on the number of genes involved, is by Fanourakis (1984). He indicated that *Ar* was independent of *df, F, ns, B, u, mc, pm, Tu,* and *D; dm* was independent of *bi, df, F, ns, ss, B, te, u, mc, Tu* and *D; bi* was independent of *cp, df, B, pm-h, te, u, mc* and *Tu; cp* was independent of *df, F, ns, ss, te, u, Tu,* and *D; F* was independent of *sf, B, pm-h, te, u, mc, Tu* and *D*; *df* was independent of *te, u, Tu*, and *D*; *ns* was independent of *B, pm-h* and *mc*; *ss* was independent of *B* and *mc*; and *B* is independent of *pm-h, te, u, Tu* and *D*.

Two other extensive studies indicated that *yc-2* was not linked with *rc, yc-1, de, bi, cr, glb, gl*, and *m*, (Whelan et al., 1975) and both *Ccu* and *Bw* were independent of *bi, gl, glb, ls, rc, sc, cr, mc, gy-1* and *gy-2* (Abul-Hayja et al., 1975). Meanwhile, white immature fruit color (*w*) was inherited independently of black spines (*B*), and locule number (*l*) (Cochran, 1938; Youngner, 1952).

Whelan (1973) found that light sensitive (*ls*) was not linked with nonbitter (*bi?*) but did not indicate which bitter gene he used. Zijlstra (1987) also determined that *bi* was independent of *cp, gl* is independent of *lh* and *ccu* is independent of *lh, ro* and *cp*.

Powdery mildew has been the subject of several linkage studies. Robinson (1978e) indicated that resistance in 'Ashley' which contains 3 recessive factors was independent of *B, l, pr, yg, fa, s*, and *H*. Kooistra (1971) found that powdery mildew resistance was not linked with *yf* or *wf* and Barham (1953) determined that the resistance genes in USDA PI 173889 were independent of *Bt*.

Like linkage data, independent assortment data may be very valuable in developing gene maps, but care must be taken when utilizing them. For example, resistance to powdery mildew was demonstrated in the previous paragraph but none of the researchers were able to identify the particular gene involved.

Literature Cited

1.   Aalders, L. E. 1959. 'Yellow Cotyledon', a new cucumber mutation. Can. J. Cytol. 1:10-12.

2.   Abul-Hayja, Z. and P. H. Williams. 1976. Inheritance of two seedling markers in cucumber. HortScience 11:145.

3.   Abul-Hayja, Z., P. H. Williams, and C. E. Peterson. 1978. Inheritance of resistance to anthracnose and target leaf spot in cucumbers. Plant Dis. Rptr. 62:43-45.

4.   Abul-Hayja, Z., P. H. Williams, and E. D. P. Whelan. 1975. Independence of scab and bacterial wilt resistance and ten seedling markers in cucumber. HortScience 10:423¬ 424.

5.   Ahnert, V., C. May, R. Gerke, and H. Kindl. 1996. Cucumber T-complex protein. Molecular cloning, bacterial expression and characterization within a 22-S cytosolic complex in cotyledons and hypocotyls. European J. of Biochemi. 235:114-119.

6.   Aleksandrov, S. V. 1952. The use of hybrid seed for increasing cucumber yields in greenhouses (in Russian). Sad. I. ogorod. 10:43-45.

7.   Andeweg, J. M. 1956. The breeding of scab-resistant frame cucumbers in the Netherlands. Euphytica 5:185-195.

8.   Andeweg, J. M. and J. W. DeBruyn. 1959. Breeding non-bitter cucumbers. Euphytica 8:13-20.

9.   Ando K, Havenga L, Grumet R. 2007. Cucumber accession PI308916, noted for compact habit and poor seedling emergence, exhibits poor apical hook formation. Cucurbit Genet Coop. 30:11-14

10.   Angelov, D. 1994. Inheritance of resistance to downy mildew, Pseudoperonospora cubensis (Berk. & Curt.) Rostow. Rep. 2nd Natl. Symp. Plant Immunity (Plovdiv) 3:99-105.

11.   Atsmon, D. and C. Tabbak. 1979. Comparative effects of gibberellin, silver nitrate and aminoethoxyvinyl-glycine on sexual tendency and ethylene evolution in the cucumber plant (*Cucumis sativus* L.). Plant and Cell Physiol. 20:1547-1555.

12.   Bailey, R. M. and I. M. Burgess. 1934. Breeding cucumbers resistant to scab. Proc. Amer. Soc. Hort. Sci. 32:474-476.

13.   Barham, W. S. 1953. The inheritance of a bitter principle in cucumbers. Proc. Amer. Soc. Hort. Sci. 62:441-442.

14.   Barnes, W. C. 1955. They both resist downy mildew: Southern cooperative trials recommend two new cucumbers. Seedsmans Digest. 14: 46-47.

15.   Barnes, W. C., C. N. Clayton, and J. M. Jenkins. 1946. The development of downy mildew-resistant cucumbers. Proc. Amer. Soc. Hort. Sci. 47: 357-60.

16.   Barnes, W. C. and W. M. Epps. 1952. Two types of anthracnose resistance in cucumbers. Plant Dis. Rptr. 36:479-480.

17.   Barnes, W.C. and W.M Epps. 1954. An unreported type of resistance to cucumber downy mildew. Plant Dis. Rptr. 38:620.Barnes, W. C. and W. M. Epps. 1955. Progress in breeding cucumbers resistant to anthracnose and downy mildew. Proc. Amer. Soc. Hort. Sci. 65: 409-15.

18.   Bressan, R. A., L. LeCureux, L. G. Wilson, P. Filner, and L. R. Baker. 1981. Inheritance of resistance to sulfur dioxide in cucumber. HortScience 16:332-333.

19.   Burnham, M., S. C. Phatak, and C. E. Peterson. 1966. Graft-aided inheritance study of a chlorophyll deficient cucumber. Proc. Amer. Soc. Hort. Sci. 89:386-389.

20.   Call, A.D. 2010. Studies on resistance to downy mildew in cucumber (Cucumis sativus L.) Caused by Pseudoperonospora cubensis. M.S. Thesis, North Carolina State Univ., Raleigh.

21.   Call, A. C. and T. C. Wehner. 2010. Search for resistance to the new race of downy mildew in cucumber, p. 112-115. Cucurbitaceae 2010 Proceeding. ASHS Press, Alexandria, V A

22.   Cantliffe, D. J. 1972. Parthenocarpy in cucumber induced by some plant growth regulating chemicals. Can. J. Plant Sci. 52:781-785.

23.   Cantliffe, D. J., R. W. Robinson, and R. S. Bastdorff. 1972. Parthenocarpy in cucumber induced by tri-iodobenzoic acid. HortScience 7:285-286.

24.   Carlsson, G. 1961. Studies of blind top shoot and its effect on the yield of greenhouse cucumbers. Acta Agr. Scand. 11:160-162.

25.   Caruth, T. F. 1975. A genetic study of the inheritance of rupturing carpel in fruit of cucumber, *Cucumis sativus* L. Ph.D. Dis., Texas A&M Univ., College Station.

26.   Chambliss, O. L. 1978. Cucumber beetle resistance in the Cucurbitaceae: Inheritance and breeding. HortScience 13:366 (Abstract).

27.   Chono, M., T. Yamauchi, S. Yamaguchi, H. Yamane, and N. Murofushi. 1996. cDNA cloning and characterization of

a gibberellin-responsive gene in hypocotyls of *Cucumis sativus* L. Plant & Cell Physiol. 37:686-691.

28. Chono, M., K. Nemoto, H. Yamane, I. Yamaguchi, and N. Murofushi. 1998. Characterization of a protein kinase gene responsive to auxin and gibberellin in cucumber hypocotyls. Plant & Cell Physiol. 39:958-967.

29. Cochran, F. D. 1938. Breeding cucumbers for resistance to downy mildew. Proc. Amer. Soc. Hort. Sci. 35:541-543.

30. Cohen, S., E. Gertman, and N. Kedar. 1971. Inheritance of resistance to melon mosaic virus in cucumbers. Phytopathology 61:253-255.

31. Cowen, N. M. and D. B. Helsel. 1983. Inheritance of 2 genes for spine color and linkages in a cucumber cross. J. Hered. 74:308-310.

32. Criswell, A.D. 2008. Screening for downy mildew resistance in cucumber. M. S. Thesis, North Carolina State Univ., Raleigh.

33. Currence. T. M. 1954. Vegetable crops breeding. Teaching manual, Univ. Minn., St. Paul (Mimeo).

34. Da Costa, C. P. and C. M. Jones. 1971a. Cucumber beetle resistance and mite susceptibility controlled by the bitter gene in *Cucumis sativus* L. Science 172:1145-1146.

35. Da Costa, C. P. and C. M. Jones. 1971b. Resistance in cucumber, *Cucumis sativus* L., to three species of cucumber beetles. HortScience 6:340-342.

36. Della Vecchia, P. T., C. E. Peterson, and J. E. Staub. 1982. Inheritance of short-day response to flowering in crosses between a *Cucumis sativus* var. *hardwickii* (R.) Alef. line and *Cucumis sativus* L. lines. Cucurbit Genet. Coop. Rpt. 5:4.

37. Delorme, V. G. R., P. F. Mcabe, D. J. Kim, and C. J. Leaver. 2000. A matrix metalloproteinase gene is expressed at the boundary of senescence and programmed cell death in cucumber. Plant Physiol. 123:917-927.

38. Denna, D. W. 1971. Expression of determinate habit in cucumbers. J. Amer. Soc. Hort. Sci. 96:277-279.

39. den Nijs, A. P. M. and I. W. Boukema. 1983. Results of linkage studies and the need for a cooperative effort to map the cucumber genome. Cucurbit Genet. Coop. Rpt. 6:22-23.

40. den Nijs, A. P. M. and I. W. Boukema. 1985. Short petiole, a useful seedling marker for genetic studies in cucumber. Cucurbit Genet. Coop. Rpt. 8:7-8.

41. den Nijs, A. P. M. and O. M. B. de Ponti. 1983. Umbrella leaf: a gene for sensitivity to low humidity in cucumber. Cucurbit Genet. Coop. Rpt. 6:24.

42. den Nijs, A. P. M. and H. O. Mackiewicz. 1980. "Divided leaf", a recessive seedling marker in cucumber. Cucurbit Genet. Coop. Rpt. 3:24.

43. de Ponti, O. M. and F. Garretsen. 1976. Inheritance of parthenocarpy in pickling cucumbers (*Cucumis sativus* L.) and linkage with other characters. Euphytica 25:633-642.

44. de Ruiter, A. C., B. J. van der Knapp, and R. W. Robinson. 1980. Rosette, a spontaneous cucumber mutant arising from cucumber-muskmelon pollen. Cucurbit Genet. Coop. Rpt. 3:4.

45. Dessert, J. M., L. R. Baker, and J. F. Fobes. 1982. Inheritance of reaction to *Pseudomonas lachrymans* in pickling cucumber. Euphytica 31:847-856.

46. Doruchowski, R. W. and E. Lakowska-Ryk. 1992. Inheritance of resistance to downy mildew (*Pseudoperonospora cubensis* Berk & Curt) in *Cucumis sativus*. Proc. 5th EUCARPIA Cucurbitaceae Symp. p. 66-69, Warsaw, Poland.

47. Fanourakis, N. E. 1984. Inheritance and linkage studies of the fruit epidermis structure and investigation of linkage relations of several traits and of meiosis in cucumber. Ph.D. Diss., Univ. of Wisconsin, Madison.

48. Fanourakis, N. E. and P. W. Simon. 1987. Analysis of genetic linkage in the cucumber. J. Hered. 78:238-242.

49. Fanourakis, N. E. and P. W. Simon. 1987. Inheritance and linkage studies of the fruit epidermis structure in cucumber. J. Hered. 78:369-371.

50. Filipecki, M. K., H. Sommer, and S. Malepszy. 1997. The MADS-box gene CUS1 is expressed during cucumber somatic embryogenesis. Plant Sci. 125:63-74.

51. Fujieda, K. and R. Akiya. 1962. Genetic study of powdery mildew resistance and spine color on fruit in cucumber. J. Jpn. Soc. Hort. Sci. 31:30-32.

52. Fujieda, K., V. Fujita, Y. Gunji, and K. Takahashi. 1982. The inheritance of plural-pistillate flowering in cucumber. J. Jap. Soc. Hort. Sci. 51:172-176.

53. Fujii, N., M. Kamada, S. Yamasaki, and H. Takahashi. 2000. Differential accumulation of *Aux/IAA* mRNA during seedling development and gravity response in cucumber (*Cucumis sativus* L.). Plant Mol. Bio. 42:731-740.

54. Galun, E. 1961. Study of the inheritance of sex expression in the cucumber. The interaction of major genes with modifying genetic and non-genetic factors. Genetica 32:134-163.

55. Ganal, M., I. Riede, And V. Hemleben. 1986. Organization and sequence analysis of two related satellite DNAs in cucumber (*Cucumis sativus* L.). J. Mol. Evol. 23:23-30.

56. Ganal, M. and V. Hemleben. 1988a. Insertion and amplification of a DNA sequence in satellite DNA of *Cucumis sativus* (cucumber). Theor. Appl. Genet. 75:357-361.

57. Ganal, M., R. Torres, and V. Hemleben. 1998b. Complex structure of the ribosomal DNA spacer of *Cucumis sativus* (cucumber). Mol. General Genet. 212:548-554.

58. George, W. L., Jr. 1970. Genetic and environmental modification of determinant plant habit in cucumbers. J. Amer. Soc. Hort. Sci. 95:583-586.

59. Goode, M. J., J. L. Bowers, and A. Bassi, Jr. 1980. Littleleaf, a new kind of pickling cucumber plant. Ark. Farm Res. 29:4.

60. Gornitskaya, I. P. 1967. A spontaneous mutant of cucumber variety Nezhinskii 12. Genetika 3(11):169.

61. Gottlieb, T. D. 1977. Evidence for duplication and divergence of the structural gene for phosphoglucoseisomerase in diploid species of *Clarkia*. Genetics 86:289-307.

62. Graham, I. A., L. M. Smith, J. W. Brown, C. J. Leaver, and S. M. Smith. 1989. The malate synthase gene of cucumber. Plant Mol. Bio. 13:673-684.

63. Graham, I. A., L. M. Smith, C. J. Leaver, and S. M. Smith. 1990. Developmental regulation of expression of the malate synthase gene in transgenic plants. Plant Mol. Bio. 15:539-549.

64. Greenler, J. M., J. S. Sloan, B. W. Schwartz, and W. M. Becker. 1989. Isolation, characterization and sequence analysis of a full-length cDNA clone encoding NADH-dependent hydroxypyruvate reductase from cucumber. Plant Mol.

Bio. 13: 139-150.

65. Greenland, A. J., M. V. Thomas, and R. W. Walde. 1987. Expression of two nuclear genes encoding chloroplast proteins during early development of cucumber seedlings. Planta 170:99-110.

66. Grimbly, P. E. 1980. An apetalous male sterile mutant in cucumber. Cucurbit Genet. Coop. Rpt. 3:9.

67. Groff, D. and M. L. Odland. 1963. Inheritance of closed-flower in the cucumber. J. Hered. 54:191-192.

68. Hande, S. and C. Jayabaskaran. 1997. Nucleotide sequence of a cucumber chloroplast protein tRNA. J. Biosci. 22:143-147.

69. Hande, S. and C. Jayabaskaran. 1996. Cucumber chloroplast trnL(CAA) gene: nucleotide sequence and in vivo expression analysis in etiolated cucumber seedlings treated with benzyladenine and light. Indian J. Biochem. Biophys. 33:448-454.

70. Havey, M. J., Y. H. Park, and G. Bartoszewski. 2004. The *Psm* locus controls paternal sorting of the cucumber mitochondrial genome. J. Hered. 95:492-497.

71. Hšhne, M., A. Nellen, K. Schwennesen, and H. Kindl. 1996. Lipid body lipoxygenase characterized by protein fragmentation, cDNA sequence and very early expression of the enzyme during germination of cucumber seeds. Eur. J. Biochem. 241:6-11.

72. Hutchins, A. E. 1935. The inheritance of a green flowered variation in *Cucumis sativus*. Proc. Amer. Soc. Hort. Sci. 33:513.

73. Hutchins, A. E. 1940. Inheritance in the cucumber. J. Agr. Res. 60:117-128.

74. Iezzoni, A. F. and C. E. Peterson. 1979. Linkage of bacterial wilt resistance and sex expression genes in cucumber. Cucurbit Genet. Coop. Rpt. 2:8.

75. Iezzoni, A. F. and C. E. Peterson. 1980. Linkage of bacterial wilt resistance and sex expression in cucumber. HortScience 15:257-258.

76. Iezzoni, A. F., C. E. Peterson, and G. E. Tolla. 1982. Genetic analysis of two perfect flowered mutants in cucumber. J. Amer. Soc. Hort. Sci. 107:678-681.

77. Iida, S. and E. Amano. 1990. Pollen irradiation to obtain mutants in monoecious cucumber. Gamma Field Symp. 29:95-111.

78. Iida, S. and E. Amano. 1991. Mutants induced by pollen irradiation in cucumber. Cucurbit Genet. Coop. Rpt. 14:32-33.

79. Inggamer, H. and O. M. B. de Ponti. 1980. The identity of genes for glabrousness in *Cucumis sativus* L. Cucurbit Genet. Coop. Rpt. 3:14.

80. Iwasaki, Y., M. Komano, and T. Takabe. 1995. Molecular cloning of cDNA for a 17.5-kDa polypeptide, the psaL gene product, associated with cucumber Photosystem I. Biosci. Biotechnol. Biochem. 59:1758-1760.

81. Jayabaskaran, C. and M. Puttaraju. 1993. Fractionation and identification of cytoplasmic tRNAs and structural characterization of a phenylalanine and a leucine tRNA from cucumber hypocotyls. Biochem. Mol. Biol. Int. 31:983-995.

82. Jenkins, J. M., Jr. 1946. Studies on the inheritance of downy mildew resistance. J. Hered. 37:267-276.

83. John, C. A. and J. D. Wilson. 1952. A "ginko leafed" mutation in the cucumber. J. Hered. 43:47-48.

84. Jones, R. W. 1984. Studies related to genetic salt tolerance in the cucumber, *Cucumis sativus* L. Ph.D. Diss., Texas A&M Univ., College Station.

85. Kabelka, E. and R. Grumet. 1997. Inheritance of resistance to the Moroccan watermelon mosaic virus in the cucumber line TMG-1 and cosegregation with zucchini yellow mosaic virus resistance. Euphytica 95:237-242.

86. Kabelka, E., Z. Ullah and R. Grumet. 1997. Multiple alleles for zucchini yellow mosaic virus resistance at the *zym* locus in cucumber. Theor. Appl. Genet. 95:997-1004.

87. Kamada, M., N. Fujii, S. Aizawa, S. Kamigaichi, C. Mukai, T. Shimazu, and H. Takahashi. 2000. Control of gravimorphogenesis by auxin: accumulation pattern of *CS-IAA1* mRNA in cucumber seedlings grown in space and on the ground. Planta 211:493-501.

88. Kamachi, S., H. Sekimoto, N. Kondo, and S. Sakai. 1997. Cloning of a cDNA for a 1-aminocyclopropane-1-carboxylate synthase that is expressed during development of female flowers at the apices of *Cucumis sativus* L. Plant Cell Physiol. 38:1197-1206.

89. Kater, M. M., L. Colombo, J. Franken, M. Busscher, S. Masiero, M. M. Van-Lookeren-Campagne, and G. C. Angenent. 1998. Multiple AGAMOUS homologs from cucumber and petunia differ in their ability to induce reproductive organ fate. Plant Cell 10:171-182.

90. Kauffman, C. S. and R. L. Lower. 1976. Inheritance of an extreme dwarf plant type in the cucumber. J. Amer. Soc. Hort. Sci. 101:150-151.

91. Kennard, W. C., K. Poetter, A. Dijkhuizen, V. Meglic, J. E. Staub, and M. J. Havey. 1994. Linkages among RFLP, RAPD, isozyme, disease-resistance, and morphological markers in narrow and wide crosses of cucumber. Theor. Appl. Genet. 89:42-48.

92. Kim, D. J. and S. M. Smith. 1994a. Molecular cloning of cucumber phosphoenolpyruvate carboxykinase and developmental regulation of gene expression. Plant Mol. Biol. 26:423-434.

93. Kim, D. J. and S. M. Smith. 1994b. Expression of a single gene encoding microbody NAD-malate dehydrogenase during glyoxysome and peroxisome development in cucumber. Plant Mol. Biol. 26:1833-1841.

94. Kim, D. J., S. M. Smith, and C. J. Leaver. 1997. A cDNA encoding a putative SPF1-type DNA-binding protein from cucumber. Gene 185:265-269.

95. Klosinska, U., E. U. Kozik, and T. C. Wehner. 2006. Inheritance of a new trait - twin fused fruit - in cucumber. HortScience 41: 313-314.

96. Knerr, L. D. and J. E. Staub. 1992. Inheritance and linkage relationships of isozyme loci in cucumber (*Cucumis sativus* L.). Theor. Appl. Genet. 84:217-224.

97. Knerr, L. D., V. Meglic and J. E. Staub. 1995. A fourth malate dehydrogenase (MDH) locus in cucumber. HortScience 30:118-119.

98. Kooistra, E. 1968. Powdery mildew resistance in cucumber. Euphytica 17:236-244.

99. Kooistra, E. 1971. Inheritance of flesh and skin colors in powdery mildew resistant cucumbers (*Cucumis sativus* L.). Euphytica 20:521-523.

100. Kozik, E. U. and T. C. Wehner. 2006. Inheritance of chilling resistance in cucumber seedlings. Proc. Cucurbitaceae 2006, p. 121-124 (ed. G. J. Holmes). Universal Press, Raleigh, North Carolina.

101. Kozik, E. U. and T. C. Wehner. 2008. A single dominant gene Ch for chilling resistance in cucumber seedlings. J. Amer. Soc. Hort. Sci. 133: 225-227

102. Kubicki, B. 1965. New possibilities of applying different sex types in cucumber breeding. Genetica Polonica 6:241-250.

103. Kubicki, B. 1969a. Investigations of sex determination in cucumber (Cucumis sativus L.). IV. Multiple alleles of locus Acr. Genetica Polonica 10:23-68.

104. Kubicki, B. 1969b. Investigations of sex determination in cucumber (Cucumis sativus L.). V. Genes controlling intensity of femaleness. Genetica Polonica 10:69-86.

105. Kubicki, B. 1969c. Investigations on sex determination in cucumbers (Cucumis sativus L.). VI. Androecism. Genetica Polonica 10:87-99.

106. Kubicki, B. 1969d. Investigations of sex determination in cucumber (Cucumis sativus L.). VII. Trimonoecism. Genetica Polonica 10:123-143.

107. Kubicki, B. 1974. New sex types in cucumber and their uses in breeding work. Proc. XIX Intl. Hort. Congr. 3:475-485.

108. Kubicki, B. and A. Korzeniewska. 1983. Inheritance of the presence of empty chambers in fruit as related to the other fruit characters in cucumbers (Cucumis sativus L.). Genetica Polonica 24:327-342.

109. Kubicki, B. and A. Korzeniewska. 1984. Induced mutations in cucumber (Cucumis sativus L.) III. A mutant with choripetalous flowers. Genetica Polonica 25:53-60.

110. Kubicki, B., I. Goszczycka, and A. Korzeniewska. 1984. Induced mutations in cucumber (Cucumis sativus L.) II. Mutant of gigantism. Genetica Polonica 25:41-52.

111. Kubicki, B., U. Soltysiak, and A. Korzeniewska. 1986a. Induced mutations in cucumber (Cucumis sativus L.) IV. A mutant of the bush type of growth. Genetica Polonica 27:273-287.

112. Kubicki, B., U. Soltysiak, and A. Korzeniewska. 1986b. Induced mutations in cucumber (Cucumis sativus L.) V. Compact type of growth. Genetica Polonica 27:289-298.

113. Kubo, Y., Y.B. Xue, A. Nakatsuka, F.M. Mathooko, A. Inaba, and R. Nakamura. 2000. Expression of a water stress-induced polygalacturonase gene in harvested cucumber fruit. J. Japan. Soc. Hort. Sci. 69:273-279.

114. Kuroda, H., T. Masuda, H. Ohta, Y. Shioi, and K. Takamiya. 1995. Light-enhanced gene expression of NADPH-protochlorophyllide oxidoreductase in cucumber. Biochem. Biophys. Res. Commun. 210:310-316.

115. Lane, K. P. and H. M. Munger. 1985. Linkage between Corynespora leafspot resistance and powdery mildew susceptibility in cucumber (Cucumis sativus). HortScience 20(3):593 (Abstract).

116. Lawton, K. A., J. Beck, S. Potter, E. Ward, and J. Ryals. 1994. Regulation of cucumber class III chitinase gene expression. Mol. Plant Microbe. Interact. 7:48-57.

117. Lower, R. A., J. Nienhuis, and C. H. Miller. 1982. Gene action and heterosis for yield and vegetative characteristics in a cross between a gynoecious pickling cucumber and a Cucumis sativus var. hardwickii line. J. Amer. Soc. Hort. Sci. 107:75-78.

118. Masuda, S., C. Sakuta, and S. Satoh. 1999. CDNA cloning of a novel lectin-like xylem sap protein and its root-specific expression in cucumber. Plant Cell Physiol. 40:1177-1181.

119. Mathooko, F. M., M. W. Mwaniki, A. Nakatsuka, S. Shiomi, Y. Kubo, A. Inaba, and R. Nakamura. 1999. Expression characteristics of CS-ACS1, CS-ACS2 and CS-ACS3, three members of the 1-aminocyclopropane-1-carboxylate synthase gene family in cucumber (Cucumis sativus L.) fruit under carbon dioxide stress. Plant Cell Physiol. 40:164-172.

120. Matsui, K., M. Nishioka, M. Ikeyoshi, Y. Matsumura, T. Mori, and T. Kajiwara. 1998. Cucumber root lipoxygenase can act on acyl groups in phosphatidylcholine. Biochim. Biophys. Acta 1390:8-20.

121. Matsui, K., K. Hijiya, Y. Tabuchi, and T. Kajiwara. 1999. Cucumber cotyledon lipoxygenase during postgerminative growth. Its expression and action on lipid bodies. Plant Physiol. 119:1279-1287.

122. May, C., R. Preisig-Muller, M. Hohne, P. Gnau, and H. Kindl, 1998. A phospholipase A2 is transiently synthesized during seed germination and localized to lipid bodies. Biochim. Biophys. Acta 1393:267-276.

123. Metraux, J. P., W. Burkhart, M. Moyer, S. Dincher, and W. Middlesteadt. 1989. Isolation of a complementary DNA encoding a chitinase with structural homology to a bifunctional lysozyme/chitinase. Proc. Natl. Acad. Sci. USA 86:896-900.

124. Meglic, V. and J. E. Staub. 1996. Inheritance and linkage relationships of isozyme and morphological loci in cucumber (Cucumis sativus L.). Theor. Appl. Genet. 92:865-872.

125. Miller, G. A. and W. L. George, Jr. 1979. Inheritance of dwarf determinate growth habits in cucumber. J. Amer. Soc. Hort. Sci. 104:114-117.

126. Miller, J. C., Jr. and J. E. Quisenberry. 1978. Inheritance of flower bud abortion in cucumber. HortScience 13:44-45.

127. Miyamoto, K., R. Tanaka, H. Teramoto, T. Masuda, H. Tsuji, and H. Inokuchi. 1994. Nucleotide sequences of cDNA clones encoding ferrochelatase from barley and cucumber. Plant Physiol. 105:769-770.

128. Morgens, P. H., A. M. Callahan, L. J. Dunn, and F. B. Abeles. 1990. Isolation and sequencing of cDNA clones encoding ethylene-induced putative peroxidases from cucumber cotyledons. Plant Mol. Biol. 14: 715-725.

129. Munger, H. M. and D. P. Lane. 1987. Source of combined resistance to powdery mildew and Corynespora leafspot in cucumber. Cucurbit Genet. Coop. Rpt. 10:1.

130. Munger, H. M. and R. E. Wilkinson. 1975. Scab resistance in relation to fruit length in slicing cucumbers. Veg. Imp. Nwsl. 17:2.

131. Nandgaonkar, A. K. and L. R. Baker. 1981. Inheritance of multi-pistillate flowering habit in gynoecious pickling cucumber. J. Amer. Soc. Hort. Sci. 106:755-757.

132. Nersissian, A. M., Z. B. Mehrabian, R. M. Nalbandyan, P. J. Hart, G. Frackiewicz, R. S. Czernuszewicz, C. J. Bender, J. Peisach, R. G. Herrmann, and J. S. Valentine. 1996. Cloning, expression, and spectroscopic characterization of Cucumis sativus stellacyanin in its nonglycosylated form. Protein Sci.

5:2184-2192.

133. Netzer, D., S. Niegro, and F. Galun. 1977. A dominant gene conferring resistance to Fusarium wilt in cucumber. Phytopathology 67:525-527.

134. Nuttall, W. W. and J. J. Jasmin. 1958. The inheritance of resistance to bacterial wilt (*Erwinia tracheiphila* [E. F. Sm.] Holland) in cucumber. Can. J. Plant Sci. 38:401-404.

135. Odland, M. L. and D. W. Groff. 1963a. Inheritance of crinkled-leaf cucumber. Proc. Amer. Soc. Hort. Sci. 83:536-537.

136. Odland, M. L. and D. W. Groff. 1963b. Linkage of vine type and geotropic response with sex forms in cucumber *Cucumis sativus* L. Proc. Amer. Soc. Hort. Sci. 82:358-369.

137. Ohkawa, J., N. Okada, A. Shinmyo, and M. Takano. 1989. Primary structure of cucumber (*Cucumis sativus*) ascorbate oxidase deduced from cDNA sequence: homology with blue copper proteins and tissue-specific expression. Proc. Natl. Acad. Sci. USA 86:1239-1243.

138. Ohkawa, J., A. Shinmyo, M. Kanchanapoon, N. Okada, and M. Takano. 1990. Structure and expression of the gene coding for a multicopper enzyme, ascorbate oxidase of cucumber. Ann. N Y Acad. Sci. 613:483-488.

139. Ovadis, M., M. Vishnevetsky, and A. Vainstein. 1998. Isolation and sequence analysis of a gene from *Cucumis sativus* encoding the carotenoid-associated protein CHRC. Plant Physiol. 118:1536.

140. Owens, K. W. and C. E. Peterson. 1982. Linkage of sex type, growth habit and fruit length in 2 cucumber inbred backcross populations. Cucurbit Genet. Coop. Rpt. 5:12.

141. Perl-Treves, R., A. Kahana, N. Rosenman, Y. Xiang, and L. Silberstein. 1998. Expression of multiple AGAMOUS-like genes in male and female flowers of cucumber (*Cucumis sativus* L.). Plant Cell Physiol. 39:701-710.

142. Peterson, G. C. and L. M. Pike. 1992. Inheritance of green mature seed-stage fruit color in *Cucumis sativus* L. J. Amer. Soc. Hort. Sci. 117:643-645.

143. Pierce, L. K. and T. C. Wehner. 1989. Gene list for cucumber. Cucurbit Genet. Coop. Rpt. 12:91-103.

144. Pierce, L. K. and T. C. Wehner. 1990. Review of genes and linkage groups in cucumber. HortScience 25:605-615.

145. Pike L. M. and T. F. Caruth. 1977. A genetic study of the inheritance of rupturing carpel in fruit of cucumber, *Cucumis sativus* L. HortScience 12:235 (Abstract).

146. Pike, L. M. and C. E. Peterson. 1969. Inheritance of parthenocarpy in the cucumber (*Cucumis sativus* L.). Euphytica 18:101-105.

147. Poole, C. F. 1944. Genetics of cultivated cucurbits. J. Hered. 35:122-128.

148. Preisig-Müller, R. and H. Kindl. 1992. Sequence analysis of cucumber cotyledon ribulosebisphosphate carboxylase/oxygenase activase cDNA. Biochim. Biophys. Acta 1171:205-206.

149. Preisig-Müller, R. and H. Kindl. 1993a. Thiolase mRNA translated in vitro yields a peptide with a putative N-terminal presequence. Plant Mol. Biol. 22:59-66.

150. Preisig-Müller, R. and H. Kindl. 1993b. Plant dnaJ homologue: molecular cloning, bacterial expression, and expression analysis in tissues of cucumber seedlings. Arch. Biochem. Biophys. 305:30-37.

151. Preisig-Müller, R., K. Guhnemann-Schafer, and H. Kindl. 1994. Domains of the tetrafunctional protein acting in glyoxysomal fatty acid beta-oxidation. Demonstration of epimerase and isomerase activities on a peptide lacking hydratase activity. J. Biol. Chem. 269:20475-20481.

152. Provvidenti, R. 1985. Sources of resistance to viruses in two accessions of *Cucumis sativus*. Cucurbit Genet. Coop. Rpt. 8:12.

153. Provvidenti, R. 1987. Inheritance of resistance to a strain of zucchini yellow mosaic virus in cucumber. HortScience 22:102-103.

154. Pyzenkov, V. I. and G. A. Kosareva. 1981. A spontaneous mutant of the dwarf type. Bul. Applied Bot. Plant Breeding 69:15-21.

155. Pyzhenkov, V. I. and G. A. Kosareva. 1981. Description of features and their inheritance pattern in a new cucumber mutant. Biull. Vses. Inst. Rastenievod 109:5-8 (Plant Breeding Abstracts 1983, 53:478).

156. Reynolds S. J. and S. M. Smith. 1995. The isocitrate lyase gene of cucumber: isolation, characterisation and expression in cotyledons following seed germination. Plant Mol. Biol. 27:487-497.

157. Richmond, R. C. 1972. Enzyme variability in the *Drosophila williston* group. 3. Amounts of variability in the super species *D. paulistorum*. Genetics 70:87-112.

158. Robinson, R. W. 1978a. Association of fruit and sex expression in the cucumber. Cucurbit Genet. Coop. Rpt. 1:10.

159. Robinson, R. W. 1978b. Fasciation in the cucumber. Cucurbit Genet. Coop. Rpt. 1:11a.

160. Robinson, R. W. 1978c. Gene linkage in 'Lemon' cucumber. Cucurbit Genet. Coop. Rpt. 1:12.

161. Robinson, R. W. 1978d. Linkage of male sterility and sex expression genes in cucumber. Cucurbit Genet. Coop. Rpt. 1:13.

162. Robinson, R. W. 1978e. Linkage relations of genes for tolerance to powdery mildew in cucumber. Cucurbit Genet. Coop. Rpt. 1:11b.

163. Robinson, R. W. 1978f. Pleiotropic effects of the glabrous gene of the cucumber. Cucurbit Genet. Coop. Rpt. 1:14.

164. Robinson, R. W. 1979. New genes for the Cucurbitaceae. Cucurbit Genet. Coop. Rpt. 2:49-53.

165. Robinson, R. W. 1987a. Blunt leaf apex, a cucumber mutant induced by a chemical mutagen. Cucurbit Genet. Coop. Rpt. 10:6.

166. Robinson, R. W. 1987b. Chlorosis induced in glabrous cucumber by high temperature. Cucurbit Genet. Coop. Rpt. 10:7.

167. Robinson, R. W. 1987c. Cordate, a leaf shape gene with pleiotropic effects on flower structure and insect pollination. Cucurbit Genet. Coop. Rpt. 10:8.

168. Robinson, R. W. 1987d. Independence of *gl* and *yc*. Cucurbit Genet. Coop. Rpt. 10:11.

169. Robinson, R. W. 1987e. Inheritance of opposite leaf arrangement in *Cucumis sativus* L. Cucurbit Genet. Coop. Rpt. 10:10.

170. Robinson, R. W. and W. Mishanec. 1964. A radiation-induced seedling marker gene for cucumbers. Veg. Imp. Nwsl. 6:2.

171. Robinson, R. W. and W. Mishanec. 1965. A new dwarf cu-

cumber. Veg. Imp. Nwsl. 7:23.

172. Robinson, R. W. and W. Mishanec. 1967. Male sterility in the cucumber. Veg. Imp. Nwsl. 9:2.

173. Robinson, R. W., H. M. Munger, T. W. Whitaker, and G. W. Bohn. 1976. Genes of the Cucurbitaceae. HortScience 11:554-568.

174. Robinson, R. W. and J. W. Shail. 1981. A cucumber mutant with increased hypocotyl and internode length. Cucurbit Genet. Coop. Rpt. 4:19-20.

175. Robinson, R. W., T. C. Wehner, J. D. McCreight, W. R. Henderson, and C. A, John. 1982. Update of the cucurbit gene list and nomenclature rules. Cucurbit Genet. Coop. Rpt. 5:62-66.

176. Robinson, R. W. and T. W. Whitaker. 1974. Cucumis, p. 145-150. In: R. C. King (ed), Handbook of Genetics. vol. 2. Plenum, New York.

177. Rosa, J. T. 1928. The inheritance of flower types in Cucumis and Citrullus. Hilgardia 3:233-250.

178. Rowe, J. T. and J. L. Bowers. 1965. The inheritance and potential of an irradiation induced tendrilless character in cucumbers. Proc. Amer. Soc. Hort. Sci. 86:436-441.

179. Rucinska, M., K. Niemirowicz-Szczytt, and A. Korzeniewska. 1991. A cucumber (Cucumis sativus L.) mutant with yellow stem and leaf petioles. Cucurbit Genet. Coop. Rpt. 14:8-9.

180. Rucinska, M., K. Niemirowicz-Szczytt, and A. Korzeniewska. 1992a. Cucumber (Cucumis sativus L.) induced mutations. II. A second short petiole mutant. Cucurbit Genet. Coop. Rpt. 15:33-34.

181. Rucinska, M., K. Niemirowicz-Szczytt, and A. Korzeniewska. 1992b. Cucumber (Cucumis sativus L.) induced mutations. III and IV. Divided and gingko leaves. Proc. 5th EUCARPIA Cucurbitaceae Symp. p. 66-69, Warsaw, Poland.

182. Sakuta, C., A. Oda, S. Yamakawa, and S. Satoh. 1998. Root-specific expression of genes for novel glycine-rich proteins cloned by use of an antiserum against xylem sap proteins of cucumber. Plant Cell Physiol. 39:1330-1336.

183. Sano, S. and K. Asada. 1994. cDNA cloning of monodehydroascorbate radical reductase from cucumber: a high degree of homology in terms of amino acid sequence between this enzyme and bacterial flavoenzymes. Plant Cell Physiol. 35:425-437.

184. Schwartz, B. W., Sloan, J. S. and Becker, W. M. (1991). Characterization of genes encoding hydroxypyruvate reductase in cucumber. Plant Molecular Biology, 17, 941-947.

185. Shanklin, J. and Somerville, C. (1991). Stearoyl-acyl-carrier-protein desaturase from higher plants is structurally unrelated to the animal and fungal homologs Proceedings of the National Academy of Sciences of the United States of America, 88, 2510-2514.

186. Shanmugasundarum, S. and P. H. Williams. 1971. A Cotyledon marker gene in cucumbers. Veg. Imp. Nwsl. 13:4.

187. Shanmugasundarum, S., P. H. Williams, and C. E. Peterson. 1971. A recessive cotyledon marker gene in cucumber with pleiotropic effects. HortScience 7:555-556.

188. Shanmugasundarum, S., P. H. Williams, and C. E. Peterson. 1971a. Inheritance of fruit spine color in cucumber. HortScience 6:213-214.

189. Shanmugasundarum, S., P. H. Williams, and C. E. Peterson. 1971b. Inheritance of resistance to powdery mildew in cucumber. Phytopathology 61:1218-1221.

190. Shanmugasundarum, S., P. H. Williams, and C. E. Peterson. 1972. A recessive cotyledon marker gene in cucumber with pleiotropic effects. HortScience 7:555-556.

191. Shcherban, T. Y., J. Shi, D. M. Durachko, M. J. Guiltinan, S. J. McQueen-Mason, M. Shieh, and D. J. Cosgrove. 1995. Molecular cloning and sequence analysis of expansins--a highly conserved, multigene family of proteins that mediate cell wall extension in plants. Proc. Natl. Acad. Sci. USA 92:9245-9249.

192. Shifriss, O. 1950. Spontaneous mutations in the American varieties of Cucumis sativus L. Proc. Amer. Soc. Hort. Sci. 55:351-357.

193. Shifriss, O. 1961. Sex control in cucumbers. J. Hered. 52:5-12.

194. Shifriss, O. and W. L. George, Jr. 1965. Delayed germination and flowering in cucumbers. Nature (London) 506:424-425.

195. Shifriss, O., C. H. Myers, and C. Chupp. 1942. Resistance to mosaic virus in cucumber. Phytopathology 32:773-784.

196. Shimizu, S., K. Kanazawa, and A. Kato. 1963. Studies on the breeding of cucumber for resistance to downy mildew. Part 2. Difference of resistance to downy mildew among the cucumber varieties and the utility of the cucumber variety resistance to downy mildew (in Japanese). Bul. Hort. Res. Sta. Jpn. Ser. A, No. 2:80-81.

197. Shimojima, M., H. Ohta, A. Iwamatsu, T. Masuda, Y. Shioi, and K. Takamiya. 1997. Cloning of the gene for monogalactosyldiacylglycerol synthase and its evolutionary origin. Proc. Natl. Acad. Sci. USA 94:333-337.

198. Shiomi, S., M. Yamamoto, T. Ono, K. Kakiuchi, J. Nakamoto, A. Makatsuka, Y. Kudo, R. Nakamura, A. Inaba, and H. Imaseki. 1998. cDNA cloning of ACC synthase and ACC oxidase genes in cucumber fruit and their differential expression by wounding and auxin. J. Japan. Soc. Hort. Sci. 67:685-692.

199. Smith, S. M. and C. J. Leaver. 1986. Glyoxysomal malate synthase of cucumber: molecular cloning of a cDNA and regulation of enzyme synthesis during germination. Plant Physiol. 81:762-767.

200. Soans, A. B., D. Pimentel, and J. S. Soans. 1973. Cucumber resistance to the two-spotted spider mite. J. Econ. Entomol. 66:380-382.

201. Soltysiak, U. and B. Kubicki. 1988. Induced mutations in the cucumber (Cucumis sativus L.). VII. Short hypocotyl mutant. Genetica Polonica 29:315-321.

202. Soltysiak, U., B. Kubicki, and A. Korzeniewska. 1986. Induced mutations in cucumber (Cucumis sativus L.). VI. Determinate type of growth. Genetica Polonica 27:299-308.

203. Staub, J. E., R. S. Kupper, D. Schuman, T. C. Wehner, and B. May. 1985. Electrophoretic variation and enzyme storage stability in cucumber. J. Amer. Soc. Hort. Sci. 110:426-431.

204. Strong, W. J. 1931. Breeding experiments with the cucumber (Cucumis sativus L.). Sci. Agr. 11:333-346.

205. Suyama, T., Yamada, K., Mori, H., Takeno, K. and Yamaki,

S. (1999). Cloning cDNAs for genes preferentially expressed during fruit growth in cucumber. *Journal of American Society for Horticultural Science*, 124, 136-139.

206. Tanaka, R., K. Yoshida, T. Nakayashiki, T. Masuda, H. Tsuji, H. Inokuchi, and A. Tanaka. 1996. Differential expression of two hemA mRNAs encoding glutamyl-tRNA reductase proteins in greening cucumber seedlings. Plant Physiol. 110:1223-1230.

207. Teramoto, H., E. Momotani, G. Takeba, and H. Tsuji. 1994. Isolation of a cDNA clone for a cytokinin-repressed gene in excised cucumber cotyledons. Planta 193:573-579.

208. Teramoto, H., T. Toyama, G. Takeba, and T. Tsuji. 1996. Noncoding RNA for CR20, a cytokinin-repressed gene of cucumber. Plant Mol. Biol. 32:797-808.

209. Tkachenko, N. N. 1935. Preliminary results of a genetic investigation of the cucumber, *Cucumis sativus* L. Bul. Applied Plant Breeding, Ser. 2, 9:311-356.

210. Thaxton, P. M. 1974. A genetic study of the clustering characteristic of pistillate flowers in the cucumber, *Cucumis sativus* L. M.S. Thesis, Texas A&M Univ., College Station.

211. Tolla, G. E. and C. E. Peterson. 1979. Comparison of gibberellin A4/A7 and silver nitrate for induction of staminate flowers in a gynoecious cucumber line. HortScience 14:542-544.

212. Toyama, T., H. Teramoto, G. Takeba, and H. Tsuji. 1995. Cytokinin induces a rapid decrease in the levels of mRNAs for catalase, 3-hydroxy-3-methylglutaryl CoA reductase, lectin and other unidentified proteins in etiolated cotyledons of cucumber. Plant Cell Physiol. 36: 1349-1359.

213. Toyama, T., H. Teramoto, and G. Takeba. 1996. The level of mRNA transcribed from psaL, which encodes a subunit of photosystem I, is increased by cytokinin in darkness in etiolated cotyledons of cucumber. Plant Cell Physiol. 37:1038-1041.

214. Toyama, T., H. Teramoto, S. Ishiguro, A. Tanaka, K. Okada and G. Taketa. 1999. A cytokinin-repressed gene in cucumber for a bHLH protein homologue is regulated by light. Plant Cell Physiol. 40:1087-1092.

215. Trebitsh, T., J. E. Staub, and S. D. O'Neill. 1997. Identification of a 1-aminocyclopropane-1-carboxylic acid synthase gene linked to the female (F) locus that enhances female sex expression in cucumber. Plant Physiol. 113: 987-995.

216. Vakalounakis, D. J. 1992. Heart leaf, a recessive leaf shape marker in cucumber: linkage with disease resistance and other traits. J. Heredity 83:217-221.

217. Vakalounakis, D. J. 1993. Inheritance and genetic linkage of fusarium wilt (*Fusarium oxysporum* f. sp. *cucumerinum* race 1) and scab (*Cladosporium cucumerinum*) resistance genes in cucumber (*Cucumis sativus*). Annals of Applied Biology 123:359-365.

218. Vakalounakis, D. J. 1995. Inheritance and linkage of resistance in cucumber line SMR-18 to races 1 and 2 of *Fusarium oxysporum* f. sp. *cucumerinum*. Plant Pathology 44:169-172.

219. Vakalounakis, D. J. 1996. Allelism of the *Fcu-1* and *Foc* genes conferring resistance to fusarium wilt in cucumber. European J. Plant Pathology 102:855-858.

220. van Es, J. 1958. Bladruuresistantie by konkommers. Zaabelangen 12:116-117.

221. van Vliet, G. J. A. and W. D. Meysing. 1974. Inheritance of resistance to *Pseudoperonospora cubensis* Rost. in cucumber (*Cucumis sativus* L.). Euphytica 23:251-255.

222. van Vliet, G. J. A. and W. D. Meysing. 1977. Relation in the inheritance of resistance to *Pseudoperonospora cubensis* Rost. and *Sphaerotheca fuliginea* Poll. in cucumber (*Cucumis sativus* L.). Euphytica 26:793-796.

223. Vishnevetsky, M., M. Ovadis, H. Itzhaki, M. Levy, Y. Libal-Weksler, Z. Adam, and A. Vainstein. 1996. Molecular cloning of a carotenoid-associated protein from *Cucumis sativus* corollas: homologous genes involved in carotenoid sequestration in chromoplasts. Plant J. 10:1111-1118.

224. Wai, T. and R. Grumet. 1995. Inheritance of resistance to the watermelon strain of papaya ringspot virus in the cucumber line TMG-1. HortScience 30:338-340.

225. Wai, T., J. E. Staub, E. Kabelka and R. Grumet. 1997. Linkage analysis of potyvirus resistance alleles in cucumber. J. Heredity. 88: 454-458.

226. Wall, J. R. 1967. Correlated inheritance of sex expression and fruit shape in *Cucumis*. Euphytica 23:251-255.

227. Walters, S. A., T. C. Wehner, and K. R. Barker. 1996. NC-42 and NC-43: Root-knot nematode-resistant cucumber germplasm. HortScience 31:1246-1247.

228. Walters, S. A., T. C. Wehner, and K. R. Barker. 1997. A single recessive gene for resistance to the root-knot nematode (*Meloidogyne javanica*) in *Cucumis sativus* var. *hardwickii*. J. Heredity 88: 66-69.

229. Walters, S. A. and T. C. Wehner. 1998. Independence of the *mj* nematode resistance gene from 17 gene loci in cucumber. HortScience 33:1050-1052.

230. Wang, Y. J., R. Provvidenti, and R. W. Robinson. 1984. Inheritance of resistance to watermelon mosaic virus 1 in cucumber. HortScience 19:587-588.

231. Wang, Y. J., R. Provvidenti, and R. W. Robinson. 1984. Inheritance of resistance in cucumber to watermelon mosaic virus. Phytopathology 51:423-428.

232. Wasuwat, S. L. and J. C. Walker. 1961. Inheritance of resistance in cucumber to cucumber mosaic virus. Phytopathology 51:423-428.

233. Wehner, T. C. 1993. Gene list update for cucumber. Cucurbit Genet. Coop. Rpt. 16: 92-97.

234. Wehner, T. C., J. E. Staub, and C. E. Peterson. 1987. Inheritance of littleleaf and multi-branched plant type in cucumber. Cucurbit Genet. Coop. Rpt. 10:33.

235. Wehner, T. C. and Staub, J. E. 1997. Gene list for cucumber. *Cucurbit Genetics Cooperative Report*, 20:66-88.

236. Wehner, T. C., J. S. Liu, and J. E. Staub. 1998a. Two-gene interaction and linkage for bitterfree foliage in cucumber. J. Amer. Soc. Hort. Sci. 123: 401-403.

237. Wehner, T. C., J. E. Staub, and J. S. Liu. 1998b. A recessive gene for revolute cotyledons in cucumber. J. Hered. 89: 86-87.

238. Wellington, R. 1913. Mendelian inheritance of epidermal characters in the fruit of *Cucumis sativus*. Science 38:61.

239. Wellington, R. and L. R. Hawthorn. 1928. A parthenocarpic hybrid derived from a cross between and English forcing cucumber and the Arlington White Spine. Proc. Amer. Soc. Hort. Sci. 25:97-100.

240. Whelan, E. D. P. 1971. Golden cotyledon: a radiation-induced mutant in cucumber. HortScience 6:343 (abstract).

241. Whelan, E. D. P. 1972a. A cytogenetic study of a radiation-induced male sterile mutant of cucumber. J. Amer. Soc. Hort. Sci. 97:506-509.

242. Whelan, E. D. P. 1972b. Inheritance of a radiation-induced light sensitive mutant of cucumber. J. Amer. Soc. Hort. Sci. 97:765-767.

243. Whelan, E. D. P. 1973. Inheritance and linkage relationship of two radiation-induced seedling mutants of cucumber. Can. J. Genet. Cytol. 15:597-603.

244. Whelan, E. D. P. 1974. Linkage of male sterility, glabrate and determinate plant habit in cucumber. HortScience 9:576-577.

245. Whelan, E. D. P. and B. B. Chubey. 1973. Chlorophyll content of new cotyledon mutants of cucumber. HortScience 10:267-269.

246. Whelan, E. D. P., P. H. Williams, and A. Abul-Hayja. 1975. The inheritance of two induced cotyledon mutants of cucumber. HortScience 10:267-269.

247. Wilson, J. M. 1968. The relationship between scab resistance and fruit length in cucumber, *Cucumis sativus* L. M.S. Thesis, Cornell Univ., Ithaca, NY.

248. Winnik, A. G. and L. F. Vetusnjak. 1952. Intervarietal hybridization of cucumber. Agrobiologia 4:125-131.

249. Xie, J. and T. C. Wehner. 2001. Gene list 2001 for cucumber. Cucurbit Genet. Coop. Rpt. 24: 110-136.

250. Yamasaki, S., N. Fujii, and H. Takahashi. 2000. The ethylene-regulated expression of CS-ETR2 and CS-ERS genes in cucumber plants and their possible involvement with sex expression in flowers. Plant Cell Physiol. 41:608-616.

251. Youngner, V. B. 1952. A study of the inheritance of several characters in the cucumber. Ph.D. Diss. Univ. of Minnesota, St. Paul.

252. Zhang, Q., A. C. Gabert, and J. R. Baggett. 1994. Characterizing a cucumber pollen sterile mutant: inheritance, allelism, and response to chemical and environmental factors. J. Amer. Soc. Hort. Sci. 119:804-807.

253. Zentgraf, U., M. Ganal, and V. Hemleben. 1990. Length heterogeneity of the rRNA precursor in cucumber (*Cucumis sativus*). Plant Mol. Bio. 15:465-474.

254. Zijlstra, S. 1987. Further linkage studies in *Cucumis sativus* L. Cucurbit Genet. Coop. Rpt. 10:39.

255. Zijlstra, S. and A. P. M. den Nijs. 1986. Further results of linkage studies in *Cucumis sativus* L. Cucurbit Genet. Coop. Rpt. 9:55.

Table 1. The qualitative genes of cucumber

Gene	Synonym	Character	References	Supplemental references	Available
a	-	*androecious.* Produces primarily staminate flowers if recessive for *F. A* from MSU 713-5 and Gy 14; *a* from An-11 and An-314, two selections from 'E-e-szan' of China.	Kubicki, 1969		P
Ak-2	-	*Adenylate kinase* (E.C. # 2.7.4.3). Isozyme variant found segregating in PI 339247, and 271754; 2 alleles observed.	Meglic and Staub, 1996		P
Ak-3	-	*Adenylate kinase* (E.C. # 2.7.4.3). Isozyme variant found segregating in PI 113334, 183967, and 285603; 2 alleles observed.	Meglic and Staub, 1996		P
al	-	*albino cotyledons.* White cotyledons and slightly light green hypocotyl; dying before first true leaf stage. Wild type Al from 'Nishiki-suyo'; *al* from M_2 line from pollen irradiation.	Iida and Amano, 1990, 1991		?
ap	-	*apetalous.* Male-sterile. Anthers become sepal-like. *Ap* from 'Butcher's Disease Resisting'; *ap* from 'Butcher's Disease Resisting Mutant'.	Grimbly, 1980		L
Ar	-	*Anthracnose resistance.* One of several genes for resistance to *Colletotrichum lagenarium. Ar* from PI 175111, PI 175120, PI 179676, PI 183308, PI 183445; *ar* from 'Palmetto' and 'Santee'.	Barnes and Epps, 1952		P
B	-	*Black or brown spines.* Dominant to white spines on fruit. Completely linked or pleiotropic with heavy netting of fruit (*H*) and red mature fruit color (*R*).	Strong, 1931; Tkachenko, 1935; Wellington, 1913	Cochran, 1938; Fujieda and Akiya, 1962; Hutchins, 1940; Jenkins, 1946; Youngner, 1952	W
B-2	-	*Black spine-2.* Interacts with *B* to produce F_2 of 15 black: 1 white spine. *B-2* from Wis. 9362; *b-2* from PI 212233 and 'Pixie'.	Shanmugasundarum et al., 1971a		?
B-3	-	*Black spine-3.* Interacts with *B-4* to produce an F_2 of nine black: 7 white spine. *B-3* from LJ90430; *b-3* from MSU 41.	Cowen and Helsel, 1983		W
B-4	-	*Black spine-4.* Interacts conversely with *B-3. B-4* from LJ90430; *b-4* from MSU 41.	Cowen and Helsel, 1983		W
bi	-	*bitterfree.* All plant parts lacking cucurbitacins. plants with *bi* less preferred by cucumber beetles. Plants with *Bi* resistant to spider mites in most American cultivars; *bi* in most Dutch cultivars.	Andeweg and DeBruyn, 1959	Cantliffe, 1972; Da Costa and Jones, 1971a, 1971b; Soans et al., 1973	W

bi-2	-	*bitterfree-2.* Leaves lacking cucurbitacins; bi-*2* from NCG-093 (short petiole mutant).	Wehner et al., 1998a		W
bl	*t*	*blind.* Terminal bud lacking after temperature shock. *bl* from 'Hunderup' and inbred HP3.	Carlsson, 1961.		L
bla	-	*blunt leaf.* Leaves have obtuse apices and reduced lobing and serration. *bla* from a mutant of 'Wis. SMR 18'.	Robinson, 1987a		W
Bt	-	*Bitter fruit.* Fruit with extreme bitter flavor. *Bt* from PI 173889 (Wild Hanzil Medicinal Cucumber).	Barham, 1953		W
Bu	-	*bush.* Shortened internodes. *bu* from 'KapAhk 1'.	Pyzenkov and Kosareva, 1981		L
Bw	-	*Bacterial wilt resistance.* Resistance to *Erwinia tracheiphila. Bw* from PI 200818; *bw* from 'Marketer'.	Nutall and Jasmin, 1958	Robinson and Whitaker, 1974	W
by	*bu*	*bushy.* Short internodes; normal seed viability. Wild type *By* from 'Borszczagowski'; *by* from induced mutation of 'Borszczagowski'. Linked with *F* and *gy*, not with *B* or *bi*.	Kubicki et al., 1986a		?
c	-	*cream mature fruit color.* Interaction with *R* is evident in the F_2 ratio of 9 red (*RC*) : 3 orange (*Rc*) : 3 yellow (*rC*) : 1 cream (*rc*).	Hutchins, 1940		L
Cca	-	*Corynespora cassiicola resistance.* Resistance to target leaf spot; dominant to susceptibility. *Cca* from Royal Sluis Hybrid 72502; *cca* from *Gy 3*.	Abul-Hayja et al., 1975		W
Ccu	-	*Cladosporium cucumerinum resistance.* Resistance to scab. *Ccu* from line 127.31, a selfed progeny of 'Longfellow'; *ccu* from 'Davis Perfect'.	Bailey and Burgess, 1934	Abul-Hayja and Williams, 1976; Abul-Hayja et al, 1975;	W
cd	-	*chlorophyll deficient.* Seedling normal at first, later becoming a light green; lethal unless grafted. *cd* from a mutant selection of backcross of MSU 713-5 x 'Midget' F1 to 'Midget'.	Burnham, et al., 1966		L
Ch		*Seedling chilling resistance. Ch* from inbred NC-76. Originally from PI 246930	Kozik and Wehner, 2006, 2008		W
chp	-	*choripetalous.* Small first true leaf; choripetalous flowers; glossy ovary; small fruits; few seeds. Wild type *Chp* from 'Borszczagowski'; chp from chemically induced mutation.	Kubicki and Korzeniewska, 1984		?
cl	-	*closed flower.* Staminate and pistillate flowers do not open; male-sterile (nonfertile pollen).	Groff and Odland, 1963		W
cla	-	*Colletotrichum lagenarium resistance.* Resistance to race 1 of anthracnose; recessive to susceptibility. *Cla* from Wis.	Abul-Hayja et al., 1978		W

		SMR 18; *cla* from SC 19B.			
Cm	-	*Corynespora melonis resistance.* Resistance to *C. melonis* dominant to susceptibility. *Cm* from 'Spotvrie'; cm from 'Esvier'.	van Es, 1958		?
Cmv	-	*Cucumber mosaic virus resistance.* One of several genes for resistance to CMV. *Cmv* from 'Wis. SMR 12', 'Wis. SMR 15', and 'Wis. SMR 18'; *cmv* from 'National Pickling' and 'Wis. SR 6'.	Wasuwat and Walker, 1961	Shifriss et al., 1942	W
co	-	*green corolla.* Green petals that turn white with age and enlarged reproductive organs; female-sterile. *co* from a selection of 'Extra Early Prolific'.	Hutchins, 1935	Currence, 1954	L
cor-1	-	*cordate leaves-1.* Leaves are cordate. *cor-1* from 'Nezhinskii'.	Gornitskaya, 1967		L
cor-2	*cor*	*cordate leaves-2.* Leaves are nearly round with revolute margins and no serration. Insect pollination is hindered by short calyx segments that tightly clasp the corolla, preventing full opening. *cor-2* from an induced mutant of 'Lemon'.	Robinson, 1987c		?
cp	-	*compact.* Reduced internode length, poorly developed tendrils, small flowers. *cp* from PI 308916.	Kauffman and Lower, 1976	Ando et al., 2007	W
cp-2	-	*compact-2.* Short internodes; small seeds; similar to *cp*, but allelism not checked. Wild type *Cp-2* from 'Borszczagowski'; *cp-2* from induced mutation of 'Borszczagowski' called W97. Not linked with *B* or *F*; interacts with by to produce super dwarf.	Kubicki et al., 1986b		?
cr	-	*crinkled leaf.* Leaves and seed are crinkled.	Odland and Groff, 1963a		?
cs	-	*carpel splitting.* Fruits develop deep longitudinal splits. *cs* from TAMU 1043 and TAMU 72210, which are second and fifth generation selections of MSU 3249 x SC 25.	Caruth, 1975; Pike and Caruth, 1977		?
D	*g*	*Dull fruit skin.* Dull skin of American cultivars, dominant to glossy skin of most European cultivars.	Poole, 1944; Strong, 1931; Tkachenko, 1935		W
de	*I*	*determinate habit.* Short vine with stem terminating in flowers; modified by *In-de* and other genes; degree of dominance depends on gene background. *de* from Penn 76.60G[*], Minn 158.60[*], 'Hardin's PG57'[*], 'Hardin's Tree Cucumber'[*], and S₂-1 (and inbred selection from Line 541)[**].	Denna, 1971[*]; George, 1970[**]; Hutchins, 1940		W
de-2	-	*determinate-2.* Main stem growth ceases after 3 to 10 nodes, producing flowers at	Soltysiak et al., 1986		?

		the apex; smooth, fragile, dark-green leaves; similar to *de*, but not checked for allelism. Wild type *De-2* from 'Borszczagowski'; *de-2* from W-sk mutant induced by ethylene-imine from 'Borszczagowski'.			
df	-	*delayed flowering.* Flowering delayed by long photoperiod; associated with dormancy. *df* from 'Baroda' (PI 212896)[*] and PI 215589 (hardwickii)[**].	Della Vecchia et al., 1982[*]; Shifriss and George, 1965[**].		W
dl	-	*delayed growth.* Reduced growth rate; shortening of hypocotyl and first internodes. *dl* from 'Dwarf Marketmore' and 'Dwarf Tablegreen', both eriving dwarfness from 'Hardin's PG-57'.	Miller and George, 1979		W
dm-1	P	*downy mildew resistance.* One of several genes for resistance to *Pseudoperonospora cubensis*. *Dm-1* from Sluis & Groot Line 4285; *dm-1* from 'Poinsett'.	van Vliet and Meysing, 1974	Jenkins, 1946; Shimizu, 1963	W
dvl	*dl*	*divided leaf.* True leaves are partly or fully divided, often resulting in compound leaves with two to five leaflets and having incised corollas.	den Nijs and Mackiewicz, 1980		W
dvl-2	*dl-2*	*divided leaf-2.* Divided leaves after the 2nd true leaf; flower petals free; similar to *dvl*, but allelism not checked. Wild type *Dvl-2* from 'Borszczagowski'; *dvl-2* from mutant induced by ethylene-imine from 'Borszczagowski'.	Rucinska et al., 1992b		?
dw	-	*dwarf.* Short internodes. *dw* from an induced mutant of 'Lemon'.	Robinson and Mishanec, 1965		?
dwc-1	-	*dwarf cotyledons-1.* Small cotyledons; late germination; small first true leaf; died after 3rd true leaf. Wild type *Dwc-1* from 'Nishiki Suyo'; *dwc-1* from M_2 line from pollen irradiation.	Iida and Amano, 1990, 1991		?
dwc-2	-	*dwarf cotyledons-2.* Small cotyledons; late germination; small first true leaf. Wild type *Dwc-2* from 'Nishiki Suyo'; *dwc-2* from M_2 line from pollen irradiation.	Iida and Amano, 1990, 1991		?
Es-1	-	*Empty chambers-1.* Carpels of fruits separated from each other, leaving a small to large cavity in the seed cell. *Es-1* from PP-2-75; *es-1* from Gy-30-75.	Kubicki and Korzeniewska, 1983		?
Es-2	-	*Empty chambers-2.* Carpels of fruits separated from each other, leaving a small to large cavity in the seed cell. *Es-2* from PP-2-75; *es-2* from Gy-30-75.	Kubicki and Korzeniewska, 1983		?
F	*Acr*, *acr[F]*, *D*, *st*	*Female.* High degree of pistillate sex expression; interacts with *a* and *M*; strongly modified by environment and	Galun, 1961; Tkachenko, 1935	Kubicki, 1965, 1969a; Poole, 1944; Shifriss,	W

				1961	
		gene background. *F* and *f* are from 'Japanese'. Plants are andromonoecious if (*mm ff*); monoecious if (*MM ff*); gynoecious if (*MM FF*) and hermaphroditic if (*mm FF*).			
fa	-	*fasciated*. Plants have flat stems, short internodes, and rugose leaves. *fa* was from a selection of 'White Lemon'*.	Robinson, 1987b*; Shifriss, 1950		?
Fba	-	*Flower bud abortion*. Preanthesis abortion of floral buds, ranging from 10% to 100%. *fba* from MSU 0612.	Miller and Quisenberry, 1978		?
Fdp-1	-	*Fructose diphosphatase* (E.C. # 3.1.3.11). Isozyme variant found segregating in PI 192940, 169383 and 169398; 2 alleles observed.	Meglic and Staub, 1996		P
Fdp-2	-	*Fructose diphosphatase* (E.C. # 3.1.3.11). Isozyme variant found segregating in PI 137851, 164952, 113334 and 192940; 2 alleles observed.	Meglic and Staub, 1996		P
Fl	-	*Fruit length*. Expressed in an additive fashion, fruit length decreases incrementally with each copy of *fl* (H. Munger, personal communication).	Wilson, 1968		W
Foc	*Fcu-1*	*Fusarium oxysporum f. sp. cucumerinum resistance*. Resistance to fusarium wilt races 1 and 2; dominant to susceptibility. *Foc* from WIS 248; *foc* from 'Shimshon'.	Netzer et al., 1977; Vakalounakis, 1993, 1995, 1996		W
G2dh	-	*Glutamine dehydrogenase* (E.C. # 1.1.1.29). Isozyme variant found segregating in PI 285606; 5 alleles observed.	Knerr and Staub, 1992		P
g	-	*golden leaves*. Golden color of lower leaves. *G* and *g* are both from different selections of 'Nezhin'.	Tkachenko, 1935		?
gb	*n*	*gooseberry fruit*. Small, oval-shaped fruit. *gb* from the 'Klin mutant'.	Tkachenko, 1935		?
gc	-	*golden cotyledon*. Butter-colored cotyledons; seedlings die after 6 to 7 days. *gc* from a mutant of 'Burpless Hybrid'.	Whelan, 1971		W
gi	-	*ginkgo*. Leaves reduced and distorted, resembling leaves of Ginkgo; male- and female-sterile. Complicated background: It was in a segregating population whose immediate ancestors were offspring of crosses and backcrosses involving 'National Pickling', 'Chinese Long', 'Tokyo Long Green', 'Vickery', 'Early Russian', 'Ohio 31' and an unnamed white spine slicer.	John and Wilson, 1952		L
gi-2	-	*ginkgo-2*. Spatulate leaf blade with reduced lobing and altered veins; recognizable at	Rucinska et al., 1992b		?

		the 2nd true leaf stage; similar to gi, fertile instead of sterile. Wild type *Gi-2* from 'Borszczagowski'; *gi-2* from mutant in the Kubicki collection.			
gig	-	*gigantism*. First leaf larger than normal. Wild type *Gig* from 'Borszczagowski'; *gig* from chemically induced mutation.	Kubicki et al., 1984		?
gl	-	*glabrous*. Foliage lacking trichomes; fruit without spines. Iron-deficiency symptoms (chlorosis) induced by high temperature. *gl* from NCSU 75[*] and M834-6[**].	Inggamer and de Ponti, 1980[**];Robinson and Mishanec, 1964[*]	Robinson, 1987b	W
glb	-	*glabrate*. Stem and petioles glabrous, laminae slightly pubescent. *glb* from 'Burpless Hybrid'.	Whelan, 1973		W
gn	-	*green mature fruit*. Green mature fruits when *rr gngn*; cream colored when *rr GnGn*; orange when *R---*. Wild type *Gn* from 'Chipper', SMR 58 and PI 165509; *gn* from TAMU 830397.	Peterson and Pike, 1992		W
Gpi-1	-	*Glucose phosphate isomerase* (E.C. # 5.3.1.9). Isozyme variant found segregating (1 and 2) in PI 176524, 200815, 249561, 422192, 432854, 436608; 3 alleles observed.	Knerr and Staub, 1992		P
Gr-1	-	*Glutathione reductase-1* (E.C. # 1.6.4.2). Isozyme variant found segregating in PI 109275; 5 alleles observed.	Knerr and Staub, 1992		P
gy	-	*gynoecious*. Recessive gene for high degree of pistillate sex expression.	Kubicki, 1974		W
H	-	*Heavy netting of fruit*. Dominant to no netting; completely linked or pleiotropic with black spines (*B*) and red mature fruit color (*R*).	Hutchins, 1940; Tkachenko, 1935		W
hl	-	*heart leaf*. Heart shaped leaves. Wild type *Hl* from Wisconsin SMR 18; *hl* from WI 2757. Linked with *ns* and *ss* in the linkage group with *Tu-u-D-pm*.	Vakalounakis, 1992		W
hn	-	*horn like cotyledons*. Cotyledons shaped like bull horns; true leaves with round shape rather than normal lobes; circular rather than ribbed stem cross section; divided petals; spineless fruits; pollen fertile, but seed sterile. Wild type *Hn* from 'Nishiki-suyo'; *hn* from M$_2$ line from pollen irradiation.	Iida and Amano, 1990, 1991		?
hsl	-	*heart shaped leaves*. Leaves heart shaped rather than lobed; tendrils branched. Wild type *Hsl* from 'Nishiki-suyo'; *hsl* from M$_2$ line from pollen irradiation.	Iida and Amano, 1990, 1991		?
I	-	*Intensifier of P*. Modifies effect of *P* on fruit warts in *Cucumis sativus* var.	Tkachenko, 1935		?

		tuberculatus.			
Idh	-	*Isocitrate dehydrogenase* (E.C. # 1.1.1.42). Isozyme variant found segregating in PI 183967, 215589; 2 alleles observed.	Knerr and Staub, 1992		P
In-de	*In(de)*	*Intensifier of de.* Reduces internode length and branching of de plants. *In-de* and *in-de* are from different selections (S_5-1 and S_5-6, respectively) from a determinant inbred S_2-1, which is a selection of line 541.	George, 1970		?
In-F	*F*	*Intensifier of female sex expression.* Increases degree of pistillate sex expression of *F* plants. *In-F* from monoecious line 18-1; *in-F* from MSU 713-5.	Kubicki, 1969b		?
l	-	*locule number.* Many fruit locules and pentamerous androecium; five locules recessive to the normal number of three.	Youngner, 1952		W
lg-1	-	*light green cotyledons-1.* Light green cotyledons, turning dark green; light green true leaves, turning dark green; poorly developed stamens. Wild type *Lg-1* from 'Nishiki-suyo'; *lg-1* from M_2 line from pollen irradiation.	Iida and Amano, 1990, 1991		?
lg-2	-	*light green cotyledons-2.* Light green cotyledons, turning dark green (faster than lg-1; light green true leaves, turning dark green; normal stamens. Wild type *Lg-2* from 'Nishiki-suyo'; *lg-2* from M_2 line from pollen irradiation.	Iida and Amano, 1990, 1991		?
lh	-	*long hypocotyl.* As much as a 3-fold increase in hypocotyl length. *lh* from a 'Lemon' mutant.	Robinson and Shail, 1981		W
ll	-	*little leaf.* Normal-sized fruits on plants with miniature leaves and smaller stems. *ll* from Ark. 79-75.	Goode et al., 1980; Wehner et al., 1987		W
ls	-	*light sensitive.* Pale and smaller cotyledons, lethal at high light intensity. *ls* from a mutant of 'Burpless Hybrid'.	Whelan, 1972b		L
m	*a, g*	*andromonoecious.* Plants are andromonoecious if (*mm ff*); monoecious if (*MM ff*); gynoecious if (*MM FF*) and hermaphroditic if (*mm FF*). *m* from 'Lemon'*.	Rosa, 1928*; Tkachenko, 1935	Shifriss, 1961; Wall, 1967; Youngner, 1952	W
m-2	*h*	*andromonoecious-2.* Bisexual flowers with normal ovaries.	Kubicki, 1974	Iezzoni, 1982	?
Mdh-1	-	*Malate dehydrogenase-1* (E.C. # 1.1.1.37). Isozyme variant found segregating in PI 171613, 209064, 326594; 3 alleles observed.	Knerr and Staub, 1992		P
Mdh-	-	*Malate dehydrogenase-2* (E.C. # 1.1.1.37).	Knerr and Staub, 1992		P

2		Isozyme variant found segregating in PI 174164, 185690, 357835, 419214; 2 alleles observed.			
Mdh-3	-	*Malate dehydrogenase-3* (E.C. # 1.1.1.37).	Knerr et al., 1995		P
Mdh-4	*Mdh-3*	*Malate dehydrogenase-4* (E.C. # 1.1.1.37). Isozyme variant found segregating in PI 255236, 267942, 432854, 432887; 2 alleles observed.	Knerr and Staub, 1992		P
mj	-	A single recessive gene for resistance to the root-knot nematode (*Meloidogyne javanica*) from *Cucumis sativus* var. *hardwickii*, *mj* from NC-42 (LJ 90430).	Walters et al., 1996; 1997	Walters and Wehner, 1998	W
mp	*pf⁺,pfᵈ, pfᵖ*	*multi-pistillate*. Several pistillate flowers per node, recessive to single pistillate flower per node. *mp* from MSU 604G and MSU 598G.	Nandgaonkar and Baker, 1981	Fujieda et al., 1982	W
Mp-2	-	*Multi-pistillate-2*. Several pistillate flowers per node. Single dominant gene with several minor modifiers. *Mp-2* from MSU 3091-1.	Thaxton, 1974		?
Mpi-1	-	*Mannose phosphate isomerase* (E.C. # 5.3.1.8). Isozyme variant found segregating in PI 176954, and 249562; 2 alleles observed.	Meglic and Staub, 1996		P
Mpi-2	-	*Mannose phosphate isomerase* (E.C. # 5.3.1.8). Isozyme variant found segregating in PI 109275, 175692, 200815, 209064, 263049, 354952; 2 alleles observed.	Knerr and Staub, 1992		P
mpy	*mpi*	*male pygmy*. Dwarf plant with only staminate flowers. Wild type *Mpy* from Wisconsin SMR 12; *mpy* from Gnome 1, a selection of 'Rochford's Improved'.	Pyzhenkov and Kosareva, 1981		?
ms-1	-	*male sterile-1*. Staminate flowers abort before anthesis; partially female-sterile. *ms-1* from selections of 'Black Diamond' and 'A & C'.	Shifriss, 1950	Robinson and Mishanec, 1967	L
ms-2	-	*male sterile-2*. Male-sterile; pollen abortion occurs after first mitotic division of the pollen grain nucleus. *ms-2* from a mutant of 'Burpless Hybrid'.	Whelan, 1973		?
ms-2ᵖˢ⁾	-	*male sterile-2 pollen sterile*. Male-sterile; allelic to *ms-2*, but not to *ap*. *ms-2ᵖˢ⁾* from a mutant of Sunseeds 23B-X26.	Zhang et al., 1994		?
mwm	-	Moroccan watermelon mosaic virus resistance single recessive gene from Chinese cucumber cultivar 'TMG-1'	Kabelka and Grumet, 1997		W
n	-	*negative geotropic peduncle response*. Pistillate flowers grow upright; *n* from 'Lemon'; *N* produces the pendant flower	Odland and Groff, 1963b		W

		position of most cultivars.			
ns	-	*numerous spines.* Few spines on the fruit is dominant to many. *ns* from Wis. 2757.	Fanourakis, 1984; Fanourakis and Simon, 1987		W
O	*y*	*Orange-yellow corolla.* Orange-yellow dominant to light yellow. *O* and *o* are both from 'Nezhin'.	Tkachenko, 1935		?
opp	-	*opposite leaf arrangement.* Opposite leaf arrangement is recessive to alternate and has incomplete penetrance. *opp* from 'Lemon'.	Robinson, 1987e		W
P	-	*Prominent tubercles.* Prominent on yellow rind of *Cucumis sativus* var. *tuberculatus*, incompletely dominant to brown rind without tubercles. *P* from 'Klin'; *p* from 'Nezhin'.	Tkachenko, 1935		W
Pc	*P*	*Parthenocarpy.* Sets fruit without pollination. *Pc* from 'Spotvrie'[*]; *pc* from MSU 713-205*.	Pike and Peterson, 1969; Wellington and Hawthorn, 1928; Whelan, 1973	de Ponti and Garretsen, 1976	?
Pe	-	*Palisade epidermis.* Epidermal cells arranged perpendicular to the fruit surface. Wild type *Pe* from 'Wisconsin SMR 18', 'Spartan Salad' and Gy 2 compact; *pe* from WI 2757.	Fanourakis and Simon, 1987		W
Pep-gl-1	-	*Peptidase with glycyl-leucine* (E.C. # 3.4.13.11). Isozyme variant found segregating in PI 113334, 212896; 2 alleles observed.	Meglic and Staub, 1996		P
Pep-gl-2	-	*Peptidase with glycyl-leucine* (E.C. # 3.4.13.11). Isozyme variant found segregating in PI 137851, 212896; 2 alleles observed.	Meglic and Staub, 1996		P
Pep-la	-	*Peptidase with leucyl-leucine* (E.C. # 3.4.13.11). Isozyme variant found segregating in PI 169380, 175692, 263049, 289698, 354952; 5 alleles observed.	Knerr and Staub, 1992		P
Pep-pap	-	*Peptidase with phenylalanyl-L-proline* (E.C. # 3.4.13.11). Isozyme variant found segregating in PI 163213, 188749, 432861; 2 alleles observed.	Knerr and Staub, 1992		P
Per-4	-	*Peroxidase* (E.C. # 1.11.1.7). Isozyme variant found segregating in PI 215589; 2 alleles observed.	Knerr and Staub, 1992		P
Pgd-1	-	*Phosphogluconate dehydrogenase-1* (E.C. # 1.1.1.43). Isozyme variant found segregating in PI 169380, 175692, 222782; 2 alleles observed.	Knerr and Staub, 1992		P
Pgd-2	-	*Phosphogluconate dehydrogenase-2* (E.C. # 1.1.1.43). Isozyme variant found	Knerr and Staub, 1992		P

		segregating in PI 171613, 177364, 188749, 263049, 285606, 289698, 354952, 419214, 432858; 2 alleles observed.			
Pgm-1	-	*Phosphoglucomutase* (E.C. # 5.4.2.2). Isozyme variant found segregating in PI 171613, 177364, 188749, 263049, 264229, 285606, 289698, 354952; 2 alleles observed.	Knerr and Staub, 1992		P
pl	-	*pale lethal.* Slightly smaller pale-green cotyledons; lethal after 6 to 7 days. *Pl* from 'Burpless Hybrid'; *pl* from a mutant of 'Burpless Hybrid'.	Whelan, 1973		L
pm-1	-	*powdery mildew resistance-1.* Resistance to *Sphaerotheca fuliginia. pm-1* from 'Natsufushinari'.	Fujieda and Akiya, 1962; Kooistra, 1971	Shanmugasundarum et al., 1972	?
pm-2	-	*powdery mildew resistance-2.* Resistance to *Sphaerotheca fuliginia. pm-2* from 'Natsufushinari'.	Fujieda and Akiya, 1962; Kooistra, 1971	Shanmugasundarum et al., 1972	?
pm-3	-	*powdery mildew resistance-3.* Resistance to *Sphaerotheca fuliginia. pm-3* found in PI 200815 and PI 200818.	Kooistra, 1971	Shanmugasundarum et al., 1972	W
pm-h	*s, pm*	*powdery mildew resistance expressed by the hypocotyl.* Resistance to powdery mildew as noted by no fungal symptoms appearing on seedling cotyledons is recessive to susceptibility. *Pm-h* from 'Wis. SMR 18'; *pm- h* from 'Gy 2 *cp cp*', 'Spartan Salad', and Wis. 2757.	Fanourakis, 1984; Shanmugasundarum et al., 1971b		W
pr	-	*protruding ovary.* Exerted carpels. *pr* from 'Lemon'.	Youngner, 1952.		W
prsv	*wmv-1-1*	*watermelon mosaic virus 1 resistance.* Resistance to papaya ringspot virus (formerly watermelon mosaic virus 1). Wild type *Prsv* from WI 2757; *prsv* from 'Surinam'.	Wang et al., 1984		?
Prsv-2	-	Resistance to papaya ringspot virus; *Prsv-2* from TMG-1.	Wai and Grumet, 1995	Wai et al., 1997	W
psl	*pl*	*Pseudomonas lachrymans resistance.* Resistance to *Pseudomonas lachrymans* is recessive. *Psl* from 'National Pickling' and 'Wis. SMR 18'; *psl* from MSU 9402 and Gy 14.	Dessert et al., 1982		W
Psm	-	*Paternal sorting of mitochondria.* Mitochondria sorting induced by dominant gene *Psm*, found in MSC 16; *psm* from PI 401734.	Havey et al., 2004.		W
R	-	*Red mature fruit.* Interaction with *c* is evident in the F$_2$ ratio of 9 red (*R-C-*) : 3 orange (*R-cc*) : 3 yellow (*rrC-*) : 1 cream (*rrcc*); completely linked or pleiotropic with black spines (*B*) and heavy netting of	Hutchins, 1940		W

		fruit (*H*).			
rc	-	*revolute cotyledon*. Cotyledons are short, narrow, and cupped downwards; enlarged perianth. *rc* from 'Burpless Hybrid' mutant.	Whelan et al., 1975		L
rc-2	-	recessive gene for revolute cotyledons; rc-2 from NCG-0093 (short petiole mutant)	Wehner et al., 1998b		W
ro	-	*rosette*. Short internodes, muskmelon-like leaves. *ro* from 'Megurk', the result of a cross involving a mix of cucumber and muskmelon pollen.	de Ruiter et al., 1980		W
s	*f, a*	*spine size and frequency*. Many small fruit spines, characteristic of European cultivars is recessive to the few large spines of most American cultivars.	Strong, 1931; Tkachenko, 1935	Caruth, 1975; Poole, 1944	W
s-2	-	*spine-2*. Acts in duplicate recessive epistatic fashion with *s-3* to produce many small spines on the fruit. *s-2* from Gy 14; *s-2* from TAMU 72210.	Caruth, 1975		?
s-3	-	*spine-3*. Acts in duplicate recessive epistatic fashion with *s-2* to produce many small spines on the fruit. *S-3* from Gy 14; *s-3* from TAMU 72210.	Caruth, 1975		?
sa	-	*salt tolerance*. Tolerance to high salt levels is attributable to a major gene in the homozygous recessive state and may be modified by several minor genes. *Sa* from PI 177362; *sa* from PI 192940.	Jones, 1984		P
sc	*cm*	*stunted cotyledons*. Small, concavely curved cotyledons; stunted plants with cupped leaves; abnormal flowers. *Sc sc* from Wis. 9594 and 9597.	Shanmugasundarum and Williams, 1971; Shanmugasundarum et al., 1972.		W
Sd	-	*Sulfur dioxide resistance*. Less than 20% leaf damage in growth chamber. *Sd* from 'National Pickling'; *sd* from 'Chipper'.	Bressan et al., 1981		W
sh	-	*short hypocotyl*. Hypocotyl of seedlings 2/3 the length of normal. Wild type *Sh* from 'Borszczagowski'; *sh* from khp, an induced mutant from 'Borszczagowski'.	Soltysiak and Kubicki 1988		?
shl	-	*shrunken leaves*. First and 2nd true leaves smaller than normal; later leaves becoming normal; slow growth; often dying before fruit set. Wild type *Shl* from 'Nishiki-suyo'; *shl* from M_2 line from pollen irradiation.	Iida and Amano, 1990, 1991		?
Skdh	-	*Shikimate dehydrogenase* (E.C. # 1.1.1.25). Isozyme variant found segregating in PI 302443, 390952, 487424; 2 alleles observed.	Meglic and Staub, 1996		P
sp	-	*short petiole*. Leaf petioles of first nodes 20% the length of normal. *sp* from Russian mutant line 1753.	den Nijs and de Ponti, 1985		W

sp-2	-	*short petiole-2*. Leaf petioles shorter, darker green than normal at 2-leaf stage; crinkled leaves with slow development; short hypocotyl and stem; little branching. Not tested for allelism with *sp*. Wild type *Sp-2* from 'Borszczagowski'; *sp-2* from chemically induced mutation.	Rucinska et al., 1992a		?
ss	-	*small spines*. Large, coarse fruit spines is dominant to small, fine fruit spines. *Ss* from 'Spartan Salad', 'Wis. SMR 18' and 'Gy 2 *cp* cp'; *ss* from Wis. 2757.	Fanourakis, 1984; Fanourakis and Simon, 1987		W
T	-	*Tall plant*. Tall incompletely dominant to short.	Hutchins, 1940		?
td	-	*tendrilless*. Tendrils lacking; associated with misshapen ovaries and brittle leaves. *Td* from 'Southern Pickler'; *td* from a mutant of 'Southern Pickler'.	Rowe and Bowers, 1965		W
te	-	*tender skin of fruit*. Thin, tender skin of some European cultivars; recessive to thick tough skin of most American cultivars.	Poole, 1944; Strong, 1931		W
tf		*twin fused fruit*. Two fruit fused into single unit. Type line: B 5263	Klosinska et al., 2006		W
Tr	-	*Trimonoecious*. Producing staminate, perfect, and pistillate flowers in this sequence during plant development. *Tr* from Tr-12, a selection of a Japanese cultivar belonging to the Fushinari group; *tr* from H-7-25. MOA-309, MOA-303, and AH-311-3.	Kubicki, 1969d		P
Tu	-	*Tuberculate fruit*. Warty fruit characteristic of American cultivars is dominant to smooth, non-warty fruits characteristic of European cultivars.	Strong, 1931; Wellington, 1913	Andeweg, 1956; Poole, 1944	W
u	M	*uniform immature fruit color*. Uniform color of European cultivars recessive to mottled or stippled color of most American cultivars.	Strong, 1931	Andeweg, 1956	W
ul	-	*umbrella leaf*. Leaf margins turn down at low relative humidity making leaves look cupped. *ul* source unknown.	den Nijs and de Ponti, 1983		W
v	-	*virescent*. Yellow leaves becoming green.	Poole, 1944; Tkachenko, 1935		L
vvi	-	*variegated virescent*. Yellow cotyledons, becoming green; variegated leaves.	Abul-Hayja and Williams, 1976		L
w	-	*white immature fruit color*. White is recessive to green. *W* from 'Vaughan', 'Clark's Special', 'Florida Pickle' and 'National Pickling'; *w* from 'Bangalore'.	Cochran, 1938		W
wf	-	*White flesh*. Intense white flesh color is recessive to dingy white; acts with *yf* to	Kooistra, 1971		?

		produce F$_2$ of 12 white *(WfWf YfYf* or *wfwf YfYf)* : 3 yellow *(WfWf yfyf)* : 1 orange *(wfwf yfyf)*. *Wf* from EG and G6, each being dingy white *(WfWf YfYf)*: *wf* from 'NPI ' which is orange *(wfwf yfyf)*.			
wi	-	*wilty leaves*. Leaves wilting in the field, but not in shaded greenhouse; weak growth; no fruiting. Wild type *Wi* from 'Nishiki-suyo'; *wi* from M$_2$ line from pollen irradiation.	Iida and Amano, 1990, 1991		?
Wmv	-	*Watermelon mosaic virus resistance.* Resistance to strain 2 of watermelon mosaic virus. *Wmv* from 'Kyoto 3 Feet'; *wmv* from 'Beit Alpha'.	Cohen et al., 1971		P
wmv-1-1	-	*watermelon mosaic virus-1 resistance.* Resistance to strain 1 of watermelon mosaic virus by limited systemic translocation; lower leaves may show severe symptoms. *Wmv-1-1* from Wis. 2757; *wmv-1-1* from 'Surinam'.	Wang et al., 1984	Provvidenti, 1985	?
wmv-2	-	*watermelon mosaic virus resistance.* Expressed in the cotyledon and throughout the plant; *wmv-2* from TMG-1.	Wai et al., 1997		W
wmv-3	-	*watermelon mosaic virus resistance.* Expressed only in true leaves; *wmv-3* from TMG-1.	Wai et al., 1997		W
wmv-4	-	*watermelon mosaic virus resistance.* Expressed only in true leaves; *wmv-4* from TMG-1.	Wai et al., 1997		W
wy	-	*wavy rimed cotyledons*. Wavy rimed cotyledons, with white centers; true leaves normal. Wild type *Wy* from 'Nishiki-suyo'; *wy* from M$_2$ line from pollen irradiation.	Iida and Amano, 1990, 1991		?
yc-1	-	*yellow cotyledons-1*. Cotyledons yellow at first, later turning green. *yc-1* from a mutant of Ohio MR 25.	Aalders, 1959		W
yc-2	-	*yellow cotyledons-2*. Virescent cotyledons. *yc-2* from a mutant of 'Burpless Hybrid'.	Whelan and Chubey, 1973; Whelan et al., 1975		W
yf	*v*	*yellow flesh*. Interacts with *wf* to produce F$_2$ of 12 white *(Wf Yf* and *wf Yf)* : 3 yellow *(Wf yf)* : 1 orange *(wf yf)*. *Yf* from 'Natsufushinari', which has an intense white flesh *(Yf wf)*; *yf* from PI 200815 which has a yellow flesh *(yf Wf)*.	Kooistra, 1971		P
yg	*gr*	*yellow-green immature fruit color.* Recessive to dark green and epistatic to light green. *yg* from 'Lemon'.	Youngner, 1952		W
yp	-	*yellow plant*. Light yellow-green foliage; slow growth.	Abul-Hayja and Williams, 1976		?

ys	-	*yellow stem.* Yellow cotyledons, becoming cream-colored; cream-colored stem, petiole and leaf veins; short petiole; short internode. Wild type *Ys* from 'Borszczagowski'; *ys* from chemically induced mutation.	Rucinska et al., 1991		?
zym-Dina	-	zucchini yellow mosaic virus resistance; *zym- Dina* from Dina-1.	Kabelka et al., 1997	Wai et al., 1997	P
zym-TMG1	*zymv*	*zucchini yellow mosaic virus resistance.* Inheritance is incomplete, but usually inherited in a recessive fashion; source of resistance is 'TMG-1'.	Provvidenti, 1987; Kabelka et al., 1997	Wai et al., 1997	W

[z]Asterisks on cultigens and associated references indicate the source of information for each. [y] W = Mutant available through T.C. Wehner, cucumber gene curator for the Cucurbit Genetics Cooperative; P = mutants are available as standard cultivars or accessions from the Plant Introduction Collection; ? = availability not known; L = mutant has been lost.
[*] Isozyme nomenclature follows a modified form of Staub et al. (1985) previously described by Richmond (1972) and Gottlieb (1977).

Table 2. The cloned genes of cucumber and their functin. [7]

Gene accession	Tissue source	Function	Clone type	Reference
Genes involved in seed germination or seedling development				
X85013	Cotyledon cDNA library	Encoding a T-complex protein	cDNA	Ahnert et al., 1996
AJ13371	Cotyledon cDNA library	Encoding a matrix metalloproteinases	cDNA	Delorme et al., 2000
X15425	Cotyledon cDNA library	Glyoxysomal enzyme malate synthase	Genomic DNA fragment	Graham et al., 1989; 1990
X92890	Cotyledon cDNA library	Encoding a lipid body lipoxygenase	cDNA	Höhne et al., 1996
L31899	Senescing cucumber cotyledon cDNA library	Encoding an ATP-dependent phosphoenolpyruvate carboxykinase (an enzyme of the gluconeogenic pathway)	cDNA	Kim and Smith, 1994a
L31900	Cotyledon cDNA library	Encoding microbody NAD(+)-dependent malate dehydrogenase (MDH)	cDNA	Kim and Smith, 1994b
L44134	Senescing cucumber cDNA library	Encoding a putative SPF1-type DNA binding protein	cDNA	Kim et al., 1997
U25058	Cotyledons	Encoding a lipoxygenase-1 enzyme	cDNA	Matsui et al., 1995; 1999
Y12793	Cotyledon cDNA library	Encoding a patatin like protein	cDNA	May et al., 1998
X67696	Cotyledon cDNA library	Encoding the 48539 Da precursor of thiolase	cDNA	Preisig-Muller and Kindl, 1993a
X67695	Cotyledon cDNA library	Encoding homologous to the bacterial dnaJ protein	cDNA	Preisig-Muller and Kindl, 1993b
X79365	Seedling cDNA library	Encoding glyoxysomal tetrafunctional protein	cDNA	Preisig-Muller et al., 1994
X79366	Seedling cDNA library	Encoding glyoxysomal tetrafunctional protein	cDNA	Preisig-Muller et al., 1994
Z35499	Genomic library	Encoding the glyoxylate cycle enzyme isocitrate lyase	Genomic gene	Reynolds and Smith, 1995
M59858	Cotyledon cDNA library	Encoding a stearoyl-acyl-carrier-protein (ACP) desaturase	cDNA	Shanklin and Somerville, 1991
M16219	Cotyledon cDNA library	Encoding glyoxysomal malate synthase	cDNA	Smith and Leaver, 1986
Genes involved in photosynthesis and photorespirationixbties				
M16056	Cotyledon cDNA library	Encoding ribulose bisphosphate carboxylase/oxygenase	cDNA	Greenland et al., 1987
M16057	Cotyledon cDNA library	Encoding chlorophyll a/b-binding protein	cDNA	Greenland et al., 1987
M16058	Cotyledon cDNA	Encoding chlorophyll a/b-binding	cDNA	Greenland et al.,

	library	protein		1987
X14609	cotyledon cDNA library	Encoding a NADH-dependent hydroxypyruvate reductase (HPR)	cDNA	Greenler et al., 1989
Y09444	Chloroplast genomic library	tRNA gene	Chloroplast DNA fragment	Hande and Jayabaskaran, 1997
X75799	Chloroplast genomic library	Chloroplast tRNA (Leu) (cAA) gene	Genomic DNA fragment	Hande et al., 1996
D50456	Cotyledon cDNA library	Encoding 17.5-kDa polypeptide of cucumber photosystem I	cDNA	Iwasaki et al., 1995
S69988	Hypocotyls	Cytoplasmic tRNA (Phe)	cytoplasmic DNA fragment	Jayabaskaran and Puttaraju, 1993
S78381	Cotyledon cDNA library	Encoding NADPH-protochlorophyllide oxidoreductase	cDNA	Kuroda et al., 1995
D26106	Cotyledon cDNA library	Encoding ferrochelatase	cDNA	Miyamoto et al., 1994
U65511	Green peelings cDNA library	Encoding the 182 amino acid long precursor stellacyanin	cDNA	Nersissian et al., 1996
AF099501	Petal cDNA library	Encoding the carotenoid-associated protein	cDNA	Ovadis et al., 1998
X67674	Cotyledon cDNA library	Encoding ribulosebisphosphate carboxylase/oxygenase activase	cDNA	Preisig-Muller and Kindl, 1992
X58542	Cucumber genomic library	Encoding NADH-dependent hydroxypyruvate reductase	Genomic DNA fragment	Schwartz et al., 1991
U62622	Seedling cDNA library	Encoding monogalacto-syldiacylglycerol synthase	cDNA	Shimojima et al., 1997
D50407	Cotyledon cDNA library	Encoding glutamyl-tRNA reductase proteins	cDNA	Tanaka et al., 1996
D67088	Cotyledon cDNA library	Encoding glutamyl-tRNA reductase proteins	cDNA	Tanaka et al., 1996
D83007	Cotyledon cDNA library	Encoding a subunit XI (psi-L) of photosystem I	cDNA	Toyama et al., 1996

Genes expressed mainly in roots

AB025717	Root RNA	Lectin-like xylem sap protein	cDNA	Masuda et al., 1999
U36339	Root cDNA library	Encoding root lipoxygenase	cDNA	Matsui et al., 1998
AB015173	Root cDNA library	Encoding glycine-rich protein-1	cDNA	Sakuta et al., 1998
AB015174	Root cDNA library	Encoding glycine-rich protein-1	cDNA	Sakuta et al., 1998

Flower genes

AF035438	Female flower cDNA library	MADS box protein CUM1	cDNA	Kater et al., 1998
AF035439	Female flower cDNA library	MADS box protein CUM10	cDNA	Kater et al., 1998
D89732	Seedlings	Encoding 1-aminocyclo-propane- 1-carboxylate synthase	cDNA	Kamachi et al., 1997
AB003683	seedlings	Encoding 1-aminocyclo-propane- 1-carboxylate synthase	cDNA	Kamachi et al., 1997

AB003684	Seedlings	Encoding 1-aminocyclo-propane- 1-carboxylate synthase	cDNA	Kamachi et al., 1997
AB035890	Fruit RNA	Encoding polygalacturonase	cDNA	Kubo et al., 2000
AF022377	Floral buds	Encoding agamous-like putative transcription factor (CAG1) mRNA	cDNA	Perl-Treves et al., 1998
AF022378	Floral buds	Encoding agamous like putative transcription factor (CAG2) mRNA	cDNA	Perl-Treves et al., 1998
AF022379	Floral buds	Encoding agamous-like putative transcription factor (CAG3) mRNA	cDNA	Perl-Treves et al., 1998
U59813	Genomic DNA	Encoding 1-aminocyclo-propane- 1-carboxylate synthase	Genomic DNA fragment	Trebitsh et al., 1997
X95593	Corolla cDNA library	Encoding carotenoid-associated protein	cDNA	Vishnevetsky et al., 1996
AB026498	Shoot apex RNA	Ethylene-receptor-related gene	cDNA	Yamasaki et al., 2000

Genes involved in fruit development and maturation

AB010922	Fruit cDNA library	Encoding the ACC synthase	cDNA	Mathooko et al., 1999
J04494	Fruit cDNA library	Encoding an ascorbate oxidase	cDNA	Ohkawa et al., 1989; 1990
AB006803	Fruit cDNA library	Encoding ACC synthase	cDNA	Shiomi et al., 1998
AB006804	Fruit cDNA library	Encoding ACC synthase	cDNA	Shiomi et al., 1998
AB006805	Fruit cDNA library	Encoding ACC synthase	cDNA	Shiomi et al., 1998
AB006806	Fruit cDNA library	Encoding ACC oxidase	cDNA	Shiomi et al., 1998
AB006807	Fruit cDNA library	Encoding ACC oxidase	cDNA	Shiomi et al., 1998
AB008846	Pollinated fruit cDNA library	Corresponding genes preferentially expressed in the pollinated fruit	cDNA	Suyama et al., 1999
AB008847	Pollinated fruit cDNA library	Corresponding genes preferentially expressed in the pollinated fruit	cDNA	Suyama et al., 1999
AB008848	Pollinated fruit cDNA library	Corresponding genes preferentially expressed in the pollinated fruit	cDNA	Suyama et al., 1999

Genes involved in cell wall loosening and cell enlargement

AB001586	Hypocotyl RNA	Encoding homologous to serine/threonine protein kinases (for CsPK1.1)	cDNA	Chono et al., 1999
AB001587	Hypocotyl RNA	Encoding homologous to serine/threonine protein kinases (for CsPK1.2)	cDNA	Chono et al., 1999
AB001588	Hypocotyl RNA	Encoding homologous to serine/threonine protein kinases (for CsPK2.1)	cDNA	Chono et al., 1999
AB001589	Hypocotyl RNA	Encoding homologous to serine/threonine protein kinases (for CsPK2.2)	cDNA	Chono et al., 1999
AB001590	Hypocotyl RNA	Encoding homologous to serine/threonine protein kinases (for	cDNA	Chono et al., 1999

		CsPK3)		
AB001591	Hypocotyl RNA	Encoding homologous to serine/threonine protein kinases (for CsPK4.1)	cDNA	Chono et al., 1999
AB001592	Hypocotyl RNA	Encoding homologous to serine/threonine protein kinases (for CsPK4.2)	cDNA	Chono et al., 1999
AB001593	Hypocotyl RNA	Encoding homologous to serine/threonine protein kinases (for CsPK5)	cDNA	Chono et al., 1999
U30382	Hypocotyl cDNA library	Encoding expansins	cDNA	Shcherban et al., 1995
U30460	Hypocotyl cDNA library	Encoding expansins	cDNA	Shcherban et al., 1995

Genes induced or repressed by plant hormones

D49413	Hypocotyl cDNA library	Corresponding to a gibberellin-responsive gene encoding an extremely hydrophobic protein	cDNA	Chono et al., 1996
AB026821	Seedling RNA	Encoding IAA induced nuclear proteins	cDNA	Fujii et al., 2000
AB026822	Seedling RNA	Encoding IAA induced nuclear proteins	cDNA	Fujii et al., 2000
AB026823	Seedling RNA	Encoding IAA induced nuclear proteins	cDNA	Fujii et al., 2000
M32742	Cotyledon cDNA library	Encoding ethylene-induced putative peroxidases	cDNA	Morgens et al., 1990
D29684	Cotyledon cDNA library	Cytokinin-repressed gene	cDNA	Teramoto et al., 1994
D79217	Genomic library	Cytokinin-repressed gene	Genomic DNA fragment	Teramoto et al., 1996
D63451	Cotyledon cDNA library	Homologous to Arabidopsis cDNA clone 3003	cDNA	Toyama et al., 1995
D63384	Cotyledon cDNA library	Encoding catalase	cDNA	Toyama et al., 1995
D63385	Cotyledon cDNA library	Encoding catalase	cDNA	Toyama et al., 1995
D63386	Cotyledon cDNA library	Encoding catalase	cDNA	Toyama et al., 1995
D63387	Cotyledon cDNA library	Encoding lectin	cDNA	Toyama et al., 1995
D63388	Cotyledon cDNA library	Encoding 3-hydroxy-3- methylglutaryl CoA reductase	cDNA	Toyama et al., 1995
D63389	Cotyledon cDNA library	Encoding 3-hydroxy-3- methylglutaryl CoA reductase	cDNA	Toyama et al., 1995
D63388	Cotyledon cDNA library	Encoding a basic region/helix- loop-helix protein	cDNA	Toyama et al., 1999

Resistance genes

M84214	Genomic library	Encoding the acidic class III chitinase	cDNA	Lawton et al., 1994
M24365	Leave cDNA	Encoding a chitinase	cDNA	Metraux et al.,

	library				1989
D26392	Seedling cDNA library	Encoding FAD-Enzyme monodehydroascorbate (MDA) reductase	cDNA		Sano and Asada, 1994

Somatic embryo gene

X97801	Embryogenic callus cDNA library	MADS-box gene	cDNA		Filipecki et al., 1997

Repeated DNA sequences

X03768	Genomic DNA	Satellite type I	Genomic DNA fragment	Ganal et al., 1986
X03769	Genomic DNA	Satellite type II	Genomic DNA fragment	Ganal et al., 1986
X03770	Genomic DNA	Satellite type III	Genomic DNA fragment	Ganal et al., 1986
X69163	Genomic DNA	Satellite type IV	Genomic DNA fragment	Ganal et al., 1988a
X07991	rDNA	Ribosomal DNA intergenic spacer	Genomic DNA fragment	Ganal et al., 1988b
X51542	Cotyledons	Ribosomal DNA intergenic spacer	Genomic DNA fragment	Zentgraf et al., 1990

[z]Table listing includes only the sequences published in journals as well as the genebank database through the year 2001.

2011 Gene List for Melon

Catherine Dogimont

INRA, UR1052, Unité de Génétique et d'Amélioration des Fruits et Légumes, BP 94, 84143 Montfavet cedex (France)
catherine.dogimont@avignon.inra.fr

Melon (*Cucumis melo* L.) is an economically important, cross-pollinated species. Melon has 2n = 24 chromosomes and a relatively small genome (450 Mb), about three times larger than the *Arabidopsis thaliana* genome and similar to the rice genome (Arumanagathan and Earle, 1991). Melon has high intra-specific genetic variation and morphologic diversity. A great variety of genetic and molecular studies have been conducted on important agronomical traits, such as resistance to pathogens and insects, and floral and fruit traits.

The following list is the latest version of the gene list for melon. Previous gene lists were organized by Pitrat: (Pitrat, 2006), (Pitrat, 2002), (Pitrat, 1998), (Pitrat, 1994), (Pitrat, 1990), (Pitrat, 1986), (Committee, 1982). This current list has been modified from previous lists in that (1) it provides an update of the known genes and QTLs, and (2) it adds an expanded description for reported genes including sources of resistance and resistance genes, phenotypes of mutants and traits related to seeds, seedlings, plant morphology and architecture, flowers and fruits. Locations of the reported genes on the melon genetic map and linked markers useful for marker assisted selection were reported where available.

Since the first molecular marker-based melon map published in 1996 (Baudracco-Arnas and Pitrat, 1996), several genetic maps of melon have been published by several research teams, using several segregating populations. 2011 will be the year of the publication of an integrated map of melon in the framework of the International Cucurbit Genomic Initiative (Díaz et al., 2011). The integrated map has been constructed by merging data from eight independent mapping populations using genetically diverse parental lines. It spans 1150 cM distributed across the 12 melon linkage groups and comprises more than 1500 markers. Individual maps and the integrated map are available at www.icugi.org. The linkage groups were named according to Perin et al., 2002. The same nomenclature will be adopted hereafter. QTLs for 62 traits including virus resistance, fruit shape, fruit weight, sugar content have been located on this integrated map (Díaz et al., 2011).

The list of melon sequences was not updated, as melon ESTs and full cDNAs are increasing extraordinarily (www.icugi.org). A physical map of the melon genome, anchored to the genetic map, has been established (Gonzalez et al., 2010) and the complete melon genome sequence is expected by the end of the year.

Host Plant Resistance genes

Considerable attention has been given to resistance genes in melon. Genes for resistance to viruses, insects, fungi and oomycetes have been reported.

Viral Diseases

The first source for resistance to *Zucchini yellow mosaic virus* (ZYMV, Potyvirus), and for a long time the only known source, was the Indian accession PI 414723 (Pitrat et al., 1996). The resistance proved to be strain-specific and was not effective against a second pathotype of the virus. The screening of about 60 cultivars from Iran allowed the identification of three immune cultivars: Magolalena Vertbrod, Soski and Bahramabadi (Arzani and Ahoonmanesh, 2000). Among 200 melons collected in Sudan, resistance sources to ZYMV were found, mainly in wild forms (Mohamed, 1999).

Resistance to ZYMV in PI 414723 was reported to be controlled by a single dominant gene, *Zym* (Pitrat and Lecoq, 1984), which mapped to the linkage group II (former LG 4), linked to the gene *a* (*andromonoecious*) (Pitrat, 1991; Perin et al., 2002). Using the ZYMV-Nat strain (pathotype 1), Danin-Poleg et al. (1997) found that three genes were needed to confer the resistance in PI 414723 (*Zym-1*, *Zym-2* and *Zym-3*). Molecular markers linked to the resistance were identified by bulk segregant analysis (Danin-Poleg et al., 2000; Danin-Poleg et al., 2002).

A semi-dominant gene named *Fn*, independent of *Zym*, was reported to control in 'Doublon' plant wilting and necrosis after inoculation with strains of the F pathotype of ZYMV (Risser et al., 1981). The *Fn* gene was located in the linkage group V (formerly 2), at 12 cM of the *Vat* gene, conferring *Aphis gossypii* resistance (Pitrat, 1991).

Necrosis after inoculation with *Watermelon mosaic virus-Morocco* (Potyvirus) was reported to be controlled by a single dominant gene *Nm* in 'Védrantais' (*nm* in 'Ouzbèque') (Quiot-Douine et al., 1988).

Papaya ringspot virus- watermelon type (PRSV, formerly called WMV-1, Potyvirus) resistance was reported in the Indian accessions PI 180280 (Webb and Bohn, 1962; Webb, 1979), PI 180283 (Quiot et al., 1971), PI 414723 (Anagnostou and Kyle, 1996), and PI 124112 (McCreight and Fashing-Burdette, 1996) and in TGR-1551 (C-105) from Zimbabwe (Gómez-Guillamón et al., 1998). Resistance to PRSV-W is conferred by a single dominant gene, *Prv*, in PI 180280 (Webb, 1979) as well as in the lines B66-5 and WMR 29, derived from PI 180280 (Pitrat and Lecoq, 1983). An allele at the same locus was shown to incite a lethal necrotic response against French strains of PRSV-W in PI 180283 and in 72025, derived from PI 180283 (Pitrat and Lecoq, 1983). These alleles were called *Prv¹* and *Prv²*, *Prv¹* being dominant over *Prv²* (Pitrat, 1986). *Prv* has been mapped to the linkage group IX (former 5) (Pitrat, 1991; Perin et al., 2002), closely linked to the gene *Fom-1* conferring resistance to *Fusarium oxysporum* races 0 and 2 (Pitrat, 1991; Perin et al., 2002; Brotman et al., 2005). A single dominant gene, *Prv-2*, was also reported to control an incompatible reaction of PI 124112 after inoculation with PRSV (McCreight and Fashing-Burdette, 1996).

Partial resistance to *Watermelon mosaic virus* (WMV, formerly WMV-2, Potyvirus) has been reported in melon line 91213, which was selected from PI 371795 and related to PI 414723 (Moyer et al., 1985; Gray et al., 1988; Moyer, 1989), the Korean accession PI 161375 (Pitrat, 1978), and in the accessions from Iran (Latifah-1, Tashkandi and Khorasgani) and in an exotic line (Galicum) (Arzani and Ahoonmanesh, 2000). Partial resistance was reported in breeding lines obtained by successive backcrossing with selection from PI 414723; inoculated plants develop mosaic symptoms on inoculated leaves but recover from symptoms and virus infection in the youngest leaves. This partial resistance was reported to be controlled by a single dominant gene, *Wmr*, linked to the ZYMV resistance gene, *Zym* (Gilbert et al., 1994; Anagnostou et al., 2000). PI 414723 was observed to be highly susceptible to WMV after inoculation with European strains of WMV (Dogimont et al., unpublished data; Gómez-Guillamón, 1998). The accession TGR-1551 was reported to exhibit very mild symptoms and a very reduced titer of virus; this partial resistance was essentially determined by a recessive gene (Diaz-Pendon et al., 2005). Still unnamed, we propose to name it *wmr-2*.

Several cultivars originating from Asia and belonging to Oriental pickling melon (var. *conomon*) and to Oriental melon (var. *makuwa*) were reported to be highly resistant to *Cucumber mosaic virus* (CMV, Cucumovirus) (Enzie, 1943; Webb and Bohn, 1962; Risser et al., 1977; Hirai and Amemiya, 1989; Daryono et al., 2003; Diaz et al., 2003). Interestingly, some accessions from Iran were also found resistant to CMV (Arzani and Ahoonmanesh, 2000) as well as the Indian IC274014 (Dhillon et al., 2007).

Resistance to the CMV-B2 strain in the accession Yamatouri was reported to be controlled in a single dominant manner. SCAR markers linked to the gene, named *Creb-2*, were identified (Daryono et al., 2010). CMV resistance was first reported to be controlled by three recessive genes in the cross Freeman Cucumber x Noy Amid (Karchi et al., 1975). Seven QTLs were shown to be involved in resistance to three different strains of CMV in the cross Védrantais x PI 161375 (Dogimont et al., 2000); one of them, located in linkage group XII explains a large part of the resistance to the strain P9 (Dogimont et al., 2000; Essafi et al., 2009).

Among about 500 accessions tested, resistance to *Cucurbit aphid borne yellows virus* (CABYV, Polerovirus, transmitted by aphids on a persistent manner) was reported in the Indian accessions 90625 (= PI 313970), Faizabadi Phoont, PI 124112, PI 282448, and PI 414723, in the Korean accession PI 255478 and in PI 124440 from South Africa (Dogimont et al., 1996). Resistance to CABYV in PI 124112 is conferred by two independent complementary recessive genes, named *cab-1* and *cab-2* (Dogimont et al., 1997).

Partial resistance to the *Beet pseudo yellows virus* (BPYV, Crinivirus), transmitted by the whitefly *Trialeurodes vaporariorum*, was reported in a few accessions of Asian origin: Nagata Kim Makuwa, PI 161375, Cma, a wild melon collected in Northern Korea and a Spanish landrace Tendral type (Esteva et al., 1989; Nuez et al., 1991). The resistance of Cma, expressed as a delayed and milder infection, resulted from the cumulative effect of an antixenosis against the vector and resistance to the virus (Soria et al., 1996; Nuez et al., 1999). Study of segregating families under natural infection suggested that the partial resistance to BPYV in Nagata Kim Makuwa, PI 161375 and Cma was controlled by single genes, partially dominant in Nagata Kim Makuwa (gene *My*) and Cma, and partially recessive in PI 161375 (Esteva and Nuez, 1992; Nuez et al., 1999).

Resistance to *Cucurbit yellow stunting disorder virus* (CYSDV, Crinivirus) was reported in the accession TGR-1551 (C-105), from Zimbabwe, under natural infection in Spain and when subjected to controlled inoculation by viruliferous *Bemisia tabaci* and by grafting (Lopez-Sese and Gomez-Guillamon, 2000). Delayed and only slight symptoms were reported in a few accessions under natural infection conditions in the United Arab Emirates (Jupiter, Muskotaly, PI 403994) and in Spain (Hassan et al., 1991; Lopez-Sese and Gomez-Guillamon, 2000). Partial resistance to CYSDV was also reported in PI 313970 in the United States (McCreight and Wintermantel, 2008). In progenies obtained from the cross between TGR-1551 and a susceptible Spanish Piel de Sapo cultivar, the resistance was shown to be controlled by a single dominant gene, called *Cys* (Lopez-Sese and Gomez-Guillamon, 2000).

A large melon germplasm was tested for *Lettuce infectious yellows virus* (LIYV, Crinivirus) resistance in natural infection by *Bemisia tabaci* biotype A in California. A snake melon originating from Saudi Arabia was shown to exhibit very mild LIYV symptoms (McCreight, 1991; McCreight, 1992). After successive field tests and confirmation in controlled-inoculation greenhouse tests, the Indian accession PI 313970, was shown to be the most interesting source of resistance to LIYV, although an occasional plant of this accession may appear symptomatic, or have a positive ELISA for LIYV (McCreight, 1998, 2000). Resistance to LIYV in PI 313970 was shown controlled by a single dominant allele at the locus designated *Liy* (McCreight, 2000).

Melon breeding line MR-1 and PI 124112, PI 179901, PI 234607, PI 313970 and PI 414723 were reported to exhibit a partial resistance to *Cucurbit leaf crumple virus* (CuLCrV), a geminivirus transmitted by *B. tabaci* biotype B, while PI 236355 was found to be completely resistant. A single recessive gene, named *culcrv*, was reported to control resistance in PI 313970, and likely in the other resistant accessions (McCreight et al., 2008).

Gonzalez-Garza et al. reported three phenotypes when they inoculated various melon cultivars with *Melon necrotic spot virus* (MNSV, Carmovirus) (Gonzalez-Garza et al., 1979): - cultivars susceptible to systemic infection showing local lesions on the inoculated leaves followed by systemic necrotic spotting, necrotic streaks on stems, conducting finally infected plants to collapse; - cultivars showing local lesions but no systemic symptoms: 53% of the accessions tested; - immune lines remaining free of symptoms ('Improved Gulfstream', 'Perlita', 'Planters Jumbo', 'PMR 5', WMR 29 and breeding line PMR Honeydew).

Among a broad germplasm collection of melons inoculated with MNSV (532 accessions), Pitrat et al. (1996) found 7% immune accessions. The resistance was confirmed to be quite common in American cantaloupe cultivars (22 resistant accessions representing 28 % out the North American accessions tested). Some resistant accessions were found originating also from Far East and India. One recessive gene, *nsv*, controls the resistance to MNSV (Coudriet et al., 1981). First described in the American cultivar Gulfstream, the same gene was shown to be present in other American germplasm ('PMR 5', 'Planters Jumbo', VA 435) and the Asian accession PI 161375 (Coudriet et al., 1981; Pitrat, 1991). *nsv* was mapped on the linkage group XII (formely 7) (Pitrat, 1991; Baudracco-Arnas and Pitrat, 1996; Perin et al., 2002). The fine mapping and the cloning of the gene revealed that the resistance corresponds to a single nucleotide

substitution in the translation initiation factor eIF4E (Morales et al., 2002; Morales et al., 2005; Nieto et al., 2006). The same substitution was found in all the MNSV resistant accessions, suggesting that the resistance has a unique origin (Nieto et al., 2007).

Two independent dominant genes, named *Mnr-1* and *Mnr-2,* were reported to control resistance to systemic infection of MNSV in Doublon; *Mnr-1* is linked to *nsv* at 19 cM (Mallor et al., 2003).

No complete sources of resistance to *Squash mosaic virus* (SqMV, Comovirus) have been reported in melons. Tolerance was, however, observed in accessions originating from India, Afghanistan, China and Pakistan (Webb and Bohn, 1962; Provvidenti, 1989, 1993). The Korean and Chinese accessions PI 161375 and China 51 (var. *makuwa*) were described to develop delayed mosaic symptoms, reduced virus multiplication, and, interestingly, complete resistance to seed transmission of SqMV (Maestro-Tejada, 1992; Provvidenti, 1998). Resistance to seed transmission was shown to be effective against four different strains of SqMV (Provvidenti, 1998). Tolerance to foliar symptoms incited by a melon strain of SqMV was shown to be controlled by a single recessive gene in China 51, but appeared to be partially dominant against a squash pathotype of SqMV (Provvidenti, 1998). Unnamed so far, we propose to name the gene *sqmv*.

Partial resistance (restriction to the virus movement) to the SH isolate of *Cucumber green mottle mosaic virus* (CGMMV, Tobamovirus) was reported in the makuwa type Chang Bougi accession (Sugiyama et al., 2006). The resistance was controlled by two complementary, recessive genes, called *cgmmv-1* and *cgmmv-2* (Sugiyama et al., 2007).

Resistance to a complex of viruses from Egypt in PI 378062 was reported to be controlled by a single dominant gene, named *Imy*, *Interveinal mottling and yellowing resistance* (Hassan et al., 1998).

Insect resistance

Resistance to the melon-cotton aphid, *Aphis gossypii* (Homoptera: Aphididae), was first reported by Kishaba and Bohn. A dominant gene, *Ag*, was reported to control antixenosis, antibiosis under controlled no-choice tests and free-curling tolerance in LJ 90634, later called PI 414723 (Kishaba et al., 1971, 1976). Pitrat and Lecoq (1980; 1986) reported resistance in PI 161375 and in PI 414723 to several viruses when they are transmitted by *A. gossypii*. The resistance is vector-specific (only *A. gossypii*), and non-specific to viruses (CMV, ZYMV, WMV…). It co-segregates with antixenosis described previously. Resistance to viruses when they are transmitted by *A. gossypii*, is controlled by a single gene, named *Vat* (*Virus aphid transmission*). The *Vat* locus was mapped to a subtelomeric position on the linkage group V (formely 2) (Pitrat, 1991; Baudracco-Arnas and Pitrat, 1996; Brotman et al., 2002; Perin et al., 2002). A single gene was cloned by positional cloning, which confers both aphid resistance and virus resistance when they are transmitted by *A. gossypii*. The gene was shown to encode a CC-NBS-LRR protein (Dogimont et al., 2004; Pauquet et al., 2004; Dogimont et al., 2010). Four additive and two couples of epistatic QTLs affecting behaviour and biotic potential of *A. gossypii* were mapped in recombinant inbred lines derived from the cross Védrantais x PI 161375; amongst them, a major QTL, which affects both behavior and biotic potential of *A. gossypii*, corresponds to the *Vat* gene (Boissot et al., 2010).

A single dominant gene, named *Lt*, was reported to control resistance to the leafminer *Liriomyza trifolii* (Diptera : Agromyzidae) in the old French cultivar Nantais Oblong (Dogimont et al., 1999). Resistant plants exhibit fewer mines and a very high larval mortality. The resistance is inefficient towards *L. huidobrensis*.

Two complementary recessive genes (*dc-1* and *dc-2*) for resistance to the melon fruit fly, *Bractocera cucurbitae* (formely *Dacus cucurbitae*, Diptera: Tephritidae) were reported by (Sambandam and Chelliah, 1972).

A monogenic recessive resistance to cucumber beetles was reported in C922-174-B in crosses among non-bitter genotypes. The gene named *cb$_1$* (*=cb*) was shown to be efficient towards three species of Coleoptera: the banded beetle *Diabrotica balteata*, the spotted beetle *D. undecimpunctata howardi* and the stripped beetle *Acalymna vittatum* (Nugent et al., 1984). In AR Top Mark, resistance to *D. undecimpunctata howardi* was also reported to be recessive and linked to the bitterness trait, controlled by the dominant gene *Bi* (Lee and Janick, 1978) that makes the melon attractive to the spotted beetle (Nugent et al., 1984).

A dominant gene, named *Af*, was reported to control resistance to the red pumpkin beetle (*Aulacophora foveicollis*, Coleoptera: Chrysomelidae) in Casaba (Vashistha and Choudhury, 1974).

Fungal Diseases

Fusarium wilt resistance. Three genes were reported to control resistance to *Fusarium oxysporum* f.sp. *melonis*. A single dominant gene, *Fom-1*, controls resistance to *F. oxysporum* races 0 and 2; it was reported in the old French cultivar Doublon (Risser, 1973; Risser et al., 1976). *Fom-1* was mapped at a distal end of the linkage group IX (formely 5), at 2 cM from the PRSV resistance gene, *Prv^2* (Perin et al., 2002). Molecular markers for *Fom-1*, useful for marker assisted selection, were developed (Brotman

et al., 2005; Oumouloud et al., 2008; Tezuka et al., 2009; Tezuka et al., 2011). A single dominant gene, *Fom-2*, controls resistance to *F. oxysporum* races 0 and 1; it was reported in CM17187 (Risser, 1973; Risser et al., 1976). *Fom-2* was mapped to the linkage group XI (Perin et al., 2002). The gene *Fom-2* was cloned and reported to encode a NBS-LRR type R protein of the non-TIR subfamily (Joobeur et al., 2004). Molecular markers linked to *Fom-2* were developed (Zheng et al., 1999; Zheng and Wolff, 2000), but their use was not completely satisfying because of recombination (Sensoy et al., 2007). New promising molecular markers were recently designed within the gene (Wang et al., 2011). Resistance to *F. oxysporum* races 0, 1 and 2 is quite frequent (Alvarez et al., 2005). The *Fom-3* gene was reported in Perlita FR; it confers the same phenotype as *Fom-1* but segregates independently from *Fom-1* (Zink and Gubler, 1985).

Resistance to *F. oxysporum* races 0 and 2 in the Spanish var. *cantalupensis* accession Tortuga was reported to be controlled by two independent genes, one dominant and the other one recessive. The dominant likely is *Fom-1*; the recessive one was named *fom-4* (Oumouloud et al., 2010).

A major recessive gene, named *fom1.2a*, was reported to confer resistance to *F. oxysporum* race 1.2 in the Israeli breeding line BIZ. The gene was located at a distal end of the LG II (opposite to the gene *a*, *andromonoecious*) (Herman et al., 2008). A second recessive gene was previously reported to segregate in the same population (Herman and Perl-Treves, 2007). In contrast, nine QTLs were reported to control the recessive resistance to race 1.2 in the French breeding line Isabelle, derived from the Far East resistant accession Ogon 9 (Perchepied and Pitrat, 2004; Perchepied et al., 2005). The resistance of the var. *cantalupensis* accession BG-5384 from Portugal to *F. oxysporum* race 1.2 (Y pathotype) was also reported to be polygenic and recessive (Chikh-Rouhou et al., 2008; Chikh-Rouhou et al., 2010).

Powdery mildew resistance. Several dominant resistance genes to powdery mildew were reported in melon. Genetic relationship between these genes is still confused, as is the definition of powdery mildew races (McCreight, 2006; Lebeda et al., 2011). Mapping of powdery resistance genes and QTLs in several crosses has thus far located them in six distinct melon linkage groups.

Jagger et al. (1938) reported a dominant resistance gene, *Pm-1*, to powdery mildew in 'PMR 45'. In the original paper, *Pm-1* was reported to confer resistance to *Erysiphe cichoracearum* but the pathogen was misidentified and was later determined to have been *Podosphaera xanthii*. *Pm-1* likely corresponds to the gene

Pm-A, which confers resistance to *P. xanthii* race 1 in 'PMR 45', described in Epinat et al. (1993). The powdery mildew resistance gene from 'PMR 45', introgressed into a yellow-fleshed breeding line, was reported to be located in the linkage group IX, loosely linked to the PRSV resistance gene, *Prv* (Teixeira et al., 2008).

A single dominant gene, *Pm-x,* confers resistance to *P. xanthii* race 1 and 2 (at least) in PI 414723; it was located in the linkage group II, linked to the ZYMV resistance gene *Zym* and to the andromonoecious gene *a* (Pitrat, 1991; Perin et al., 2002).

A single dominant gene was reported in WMR 29, *Pm-w,* which confers resistance to *P. xanthii* races 1, 2 and 3 (Pitrat, 1991). It likely corresponds to *Pm-B* in Epinat et al. (1993). It was located in the linkage group V (formerly 2), closely linked to the *Vat* locus (Pitrat, 1991; Perin et al., 2002).

Harwood and Markarian (1968) reported two dominant genes in PI 124112, *Pm-4* and *Pm-5.* These genes may correspond to the two genes of PI 124112 reported in Perchepied et al. (2005), *PmV.1* and *PmXII.1. PmV.1* confers resistance to *P. xanthii* races 1, 2, and 3 and was located in the linkage group V, closely linked to the *Vat* locus. *Pm-XII.1* confers resistance to *P. xanthii* races 1, 2 and 5 and to *Golovinomyces cichoracearum* race 1 and was mapped to the linkage group XII. It may correspond to one of the two genes, *Pm-F* and *Pm-G,* which were reported to interact for controlling resistance to *G. cichoracearum* in PI 124112 (Epinat et al., 1993).

Two genes were reported in 'PMR 5', *Pm-1* and *Pm-2* (Bohn and Whitaker, 1964). Allelism tests clearly showed that 'PMR 5' has the same gene as 'PMR 45' to control *P. xanthii race 1. Pm-2* likely corresponds to *Pm-C,* which confers resistance to *P. xanthii* race 2 in interaction with *Pm-1* (Epinat et al., 1993). Two genes, *Pm-C (Pm-2)* and *Pm-E,* were suggested to interact in 'PMR 5' to control resistance to *G. cichoracearum* (Epinat et al., 1993). Recently, two QTLs of resistance to *P. xanthii* race 1 and N1 were located in the linkage groups II and XII in recombinant inbred lines derived from the cross PMAR No.5 x Harukei No.3 (Fukino et al., 2006; Fukino et al., 2008). These two QTLs may correspond to the same genomic regions as reported in PI 124112, with different alleles. PMAR No.5 (= AR 5) was obtained from an aphid resistant line and successive backcrosses to 'PMR 5' (McCreight et al., 1984). The results obtained by (Fukino et al., 2006; Fukino et al., 2008) suggest that powdery mildew resistance genes in PMAR No.5 may be different from those in 'PMR 5', as *Pm-1* is expected to be located in the linkage group IX (Teixeira et al., 2008).

Harwood and Markarian (1968) reported a single dominant resistance gene in PI 124111, *Pm-3.* Kenigsbuch and Cohen (1989) reported a second gene in PI 124111, *Pm-6,* independent from *Pm-3,* which confers resistance to *P. xanthii* race 2. Their relationship with the other powdery mildew resistance genes is unknown.

Resistance to the Chinese race of *P. xanthi* (with a unique reaction pattern of the commonly used melon race differentials) in the Indian accession PI 134198 was reported to be controlled by a single dominant gene, designated *Pm-8,* which was suggested to be located in the linkage group VII (Liu et al., 2010).

Resistance to *P. xanthi* races 1, 2 and 5 in TGR-1551 was reported to be controlled by two independent genes, one dominant and one recessive, each one conferring resistance to all three races (Gómez-Guillamón et al., 2006; Yuste-Lisbona et al., 2008). The dominant gene, *Pm-R,* was recently located in the linkage group V, closely linked to the *Vat* and *Pm-w* loci (Yuste-Lisbona et al., 2011); the recessive gene was putatively located in the linkage group VIII, with a LOD score lower than the threshold score (Yuste-Lisbona et al., 2011).

In the same manner, resistance to *P. xanthii* in PI 313970 or 90625 was reported to be controlled by dominant, co-dominant, and recessive genes (McCreight, 2003; McCreight and Coffey, 2007; Pitrat and Besombes, 2008). Recently, PI 313970 resistance to the race S, a new strain of *P. xanthii* from Eastern-USA, virulent on all the commonly used resistance differentials, was reported to be controlled by a single recessive gene, named *pm-S.* The relationship of *pm-S* with the previously reported resistance genes in PI 313970 is unknown (McCreight and Coffey, 2011).

Other fungi. Several genes have been described to control resistance to gummy stem blight, caused by *Didymella bryoniae* (asexual form *Mycosphaerella citrullina).* Four independent dominant genes, *Gsb-1* through *Gsb-4,* were reported to confer a high level of resistance in PI 140471, PI 157082, PI 511890, and PI 482398 (Prasad and Norton, 1967; Frantz and Jahn, 2004). In the latter accession, a recessive gene, *gsb-5,* independent from *Gsb-1, Gsb-2, Gsb-3* and *Gsb-4* was also reported (Frantz and Jahn, 2004). A single dominant gene (previously named *Mc-2),* was reported to confer a moderate level of resistance in C-1 and C-8 (Prasad and Norton, 1967); we propose to rename it *Gsb-6.*

A single dominant gene, *Ac,* was reported to control resistance to *Alternaria cucumerina* in the line MR-1 (Thomas et al., 1990). A semi-dominant gene, *Mvd,* was reported to control partial resistance to melon vine decline caused by *Acremonium cucurbitacearum* and *Monosporascus cannonballus* in the wild type accession Pat 81 (Iglesias et al., 2000).

Oomycete resistance

Sources of resistance to downy mildew caused by the oomycete *Pseudoperonospora cubensis* were reported in several Indian accessions (Dhillon et al., 2007; Fergany et al., 2011). Downy mildew resistance was reported to be controlled by two partially dominant, complementary genes, *Pc-1* and *Pc-2*, in the Indian accession PI 124111 (Cohen et al., 1985; Thomas et al., 1988; Kenigsbuch and Cohen, 1992). This accession was reported to be resistant to the six known pathotypes of downy mildew (Cohen et al., 2003). Two complementary, dominant genes (*Pc-4* and *Pc-1* or *Pc-2*) were also reported to control resistance to downy mildew in another Indian accession PI 124112 (Kenigsbuch and Cohen, 1992). Nine QTLs for resistance to *P. cubensis* were located on a melon map developed from the cross 'Védrantais' x PI 124112. Among them, a major QTL, *Pc-XII.1*, was located in the linkage group XII, closely linked to the powdery mildew resistance QTL *Pm-XII.1*, which confers resistance to *P. xanthii* races 1, 2 and 5 and *G. cichoracearum* race 1 (Perchepied et al., 2005). A single dominant gene of partial resistance, *Pc-3*, was reported in the Indian accession PI 414723 (Epinat and Pitrat, 1989). The gene *Pc-5* was reported to interact with the modifier gene, *M-Pc-*, to control downy mildew resistance in the line 5-4-2-1; in presence of *M-Pc-5*, the resistance conferred by the gene *Pc-5* is dominant, while in absence of *M-Pc-5*, the resistance is recessive (Angelov and Krasteva, 2000).

Seed and Seedling Genes

Three genes were reported to control seed coat color: the *r* gene *(red stem)* controls brown seed color and a red stem in PI 157083 (30569) (Bohn, 1968; McCreight and Bohn, 1979). The gene *Wt* (*White testa*) controls white seed testa color and is dominant to yellow or tan seed coat color (Hagiwara and Kamimura, 1936). A *White testa* gene (*Wt-2*) was also reported in PI 414723, dominant to yellow seed testa color and mapped to the linkage group IV (Périn et al., 1999). The pine-seed shape of the seeds of PI 161375 is controlled by a single recessive gene, *pin, pine-seed shape*, which was mapped to the linkage group III (Perin et al., 2002). This trait is common in melon in the pinonet Spanish type. The presence of a gelatinous sheath around the seeds (versus absence) was reported to be controlled by a single dominant gene *Gs, Gelatinous sheath* (Ganesan, 1988).

Several chlorophyll deficient mutants were reported in melon. A single recessive gene, *alb, (albino)* controls the white cotyledon, lethal mutant in Trystorp (Besombes et al., 1999). The dominant pale cotyledons mutant *Pa, Pale*, is a lethal mutation as *PaPa* are albinos and die early, while *PaPa+* have yellow cotyledons and leaves (McCreight and Bohn, 1979); *Pa* was shown to be linked to the *gl* (*glabrous*) and *r* (*red stem*) mutant genes (Pitrat, 1991). A single recessive gene, *yg* (*yellow green*), controls light green cotyledons and leaves in the line 26231 (Whitaker, 1952); it was located in the linkage group XI (former 6) (Pitrat, 1991). An allele of *yg*, first described as *lg* (*light green*) in the cross Dulce x TAM-Uvalde, was renamed *yg^w* (*yellow green Weslaco*) (Cox, 1985; Cox and Harding, 1986). A single recessive gene, *f* (*flava*), controls bronze yellow cotyledons and leaves and a reduced plant growth in the Chinese accession K2005 (Pitrat et al., 1986); it was reported to be closely linked to the *lmi* (*long main stem internode*) gene (Pitrat, 1991). A recessive mutant with a yellow ring on the cotyledons that later disappears, leaving the plants a normal green, was named *h* (*halo*) (Nugent and Hoffman, 1974); it was shown to be linked to the genes *a* (*andromonoecious*), *Pm-x* (*Powdery mildew resistance x*) and *Zym* (*Zucchini yellow mosaic virus resistance*) and was then located in the linkage group II (former 4) (Pitrat, 1991; Perin et al., 2002). Three recessive *virescent* genes *v, v-2* and *v-3* control pale cream cotyledons and hypocotyls, which turn green later; the younger leaves are light green while the older ones are normal green (Hoffman and Nugent, 1973; Dyutin, 1979; Pitrat et al., 1995); the *v-3* gene was shown to be independent to *v* (Pitrat et al., 1995). Two yellow virescent recessive mutant genes, *yv* (*yellow virescent*) and *yv-2*, were reported (allelism unknown); they control pale cotyledons, yellow green young leaves and tendrils and green older leaves, associated with a severely reduced plant growth (Zink, 1977; Pitrat et al., 1991).

The incapacity of a mutant to efficiently absorb Fe (iron) and Mn (manganese) was reported to be controlled by a recessive gene, *fe*; the mutant chlorotic leaves with green veins turn to green when iron is added to the nutrient solution (Nugent and Bhella, 1988; Jolley et al., 1991).

A single recessive gene, *ech* (*exaggerated curvature of the hook*), was shown to control the triple response of seedling germination in the dark in the presence of ethylene. Seedlings exhibit a very strong, 360° hook curvature of hypocotyls in PI 161375 (*ech*), while they exhibit a moderate, 180° curvature in 'Védrantais' and PI 414723 (*Ech*). The *ech* gene was mapped to the linkage group I (Perin et al., 2002).

Seedling bitterness due to the presence of cucurbitacins, common in honeydew or Charentais type, was shown to be dominant over non-bitter, found in most American cantaloupes, and controlled by a single gene *Bi* (*Bitter*) (Lee and Janick, 1978).

A single recessive *delayed lethal* mutant, *dlet* (formerly *dl*) was described by Zink (1990); it exhibits a reduced growth, necrotic lesions on leaves leading to premature death.

Leaf and Foliage Genes

Several genes control leaf and foliage traits in melon. Two linked dominant genes, *Ala* (*Acute leaf apex*) and *L* (*Lobed* leaves) were reported to control leaf shape in 'Main Rock' (*Ala* and *L*) crossed with 'PV Green' (*ala* and *l*) (Ganesan and Sambandam, 1985). Highly indented leaves, instead of round, are controlled by a single recessive gene, *dl* (*dissected leaf*), in URRS 4 (Dyutin, 1967). An allele of *dl* in 'Cantaloup de Bellegarde', previously described as *cut leaf*, was named *dl^v*, *dissected leaf Velich* (Velich and Fulop, 1970). A second gene *dl-2* (*dissected leaf-2*), allelism unknown, was reported as "hojas hendidas" (Esquinas Alcazar, 1975). A single dominant gene, *Sfl*, was reported to control the *subtended floral leaf* trait; the leaves bearing hermaphrodite/pistillate flowers in their axis, are sessile, small and enclosing the flowers in 'Makuwa', *Sfl*, while normal in 'Annamalai', *sfl* (Ganesan and Sambandam, 1979). Cox (1985) reported two recessive leaf mutant genes, *brittle leaf dwarf* (*bd*) and *curled leaf* (*cl*), which both affect the female fertility. Spoon-shaped leaves with upward curling of the leaf margins were reported to be controlled by a single recessive gene, named *cf* (*cochleare folium*) in a spontaneous mutant in 'Galia' (Lecouviour et al., 1995). A single recessive gene, *gl* (*glabrous*), was reported to control completely hairless plants in Arizona glA (Foster, 1963). A single recessive gene, *r* (*red stem*), controls in PI 157083 (30569) a red striped hypocotyl and red stem, especially at internodes, that is photosensitive, and reddish or tan seed coat color (Bohn, 1968; McCreight and Bohn, 1979). The genes *gl* and *r* were shown to be linked in a same linkage group (LG 3) comprising also *Pa* (*Pale*) and *ms-1* (*male sterile-1*) (McCreight, 1983; Pitrat, 1991).

Plant Architecture Genes

A single gene, recessive or incompletely dominant, called *slb*, *short lateral branching* (formerly *sb*) was suggested to control the short lateral branching trait in LB-1, a wild melon from Russia (Ohara et al., 2001). In 2008, (Fukino et al., 2008) reported two QTL for short lateral branching in a cross between a breeding line Nou 4 derived from LB-1 and the normal branching 'Earl's Favourite' (Harukei 3). The QTL mapped to LG VII and LG XI, explained, respectively, 14.8 % (The allele of Harukei 3 contributed to shorter length branches) and 42.2% (The allele of 'Nou 4' contributed to shorter length branches). A mutant lacking lateral branches, named *ab*, *abrachiate*, was reported; it produces only male flowers (Foster and Bond, 1967).

A single recessive gene, *lmi* (*long main-stem internode*), controls a long hypocotyl and a long internode length (about 20 cm) in the main stem but does not affect internode length of lateral branches in 48764 (McCreight, 1983). Three recessive genes that controlled short-internodes, *si-1*, *si-2*, *si-3* (*short internode-1, -2, -3*), were reported in three independent melon lines, UC Topmark Bush, Persia 202, and 'Maindwarf' (Denna, 1962; Paris et al., 1984; Knavel, 1990). *si-1* plants display a bush phenotype, with an extremely compact growing habit and very short (about 1 cm) internode length (Denna, 1962; Zink, 1977); *si-1* is linked to the gene *yv*, *yellow virescent* (Pitrat, 1991). Internodes of *si-2* and *si-3* plants are short but less compact than *si-1* plants. In *si-2* plants, the first internodes are short, leading to a 'bird's nest' phenotype; later internodes are not modified. In *si-3* plants, internode length is reduced at all plant development stages. Fasciation of the main stem (reaching up to 15 cm) in the Charentais type line Vilmorin 104 was controlled by a single recessive gene, named *fas*, *fasciated* (Gabillard and Pitrat, 1988).

Flower Genes

Sex determination in melon is controlled by two major genes, *a* and *g*. The *andromonoecious* gene *a* (Rosa, 1928; Poole and Grimball, 1939; Wall, 1967) controls the monoecious versus andromonoecious sex type in melon. The gene mapped to the linkage group II (Perin et al., 2002; Silberstein et al., 2003). The gene was recently cloned and was shown to encode an ACC synthase gene, *CmACS7*. The transition between monoecy and andromonoecy is conferred by a single substitution, which leads to an inactive form of this key enzyme in the ethylene biosynthesis (Boualem et al., 2008). Molecular markers linked to the gene (Noguera et al., 2005; Sinclair et al., 2006; Kim et al., 2010) and within the gene are available (Boualem et al., 2008).

The *gynoecious*, *g*, gene controls the transition of monecious plants to gynoecious plants carrying only female flowers. The gene was mapped to a distal end of the linkage group V, opposite to the *Vat* gene. Positional cloning of the gene showed that the gene *G* encodes for a transcription factor of the WIP family, *CmWIP1*. The gynoecious allele *g* corresponds to the insertion of a transposable element, which epigenetically represses the expression of *CmWIP1* (Martin et al., 2009). A third gene, named *gy* (*gynomonoecious*, previously also called *n* or *M*), interacts with *a* and *g* to produce stable gynoecious plants in the gynoecious line WI 998 (Kenigsbuch and Cohen, 1987, 1990).

Five single recessive genes of male-sterility including *ms-1* to *ms-5* were reported in melon (Bohn and Whitaker, 1949; Bohn and Principe, 1964; Lozanov,

1983; McCreight and Elmstrom, 1984; Lecouviour et al., 1990) in (Pitrat, 1991, 2002). Each of these genes displays a unique phenotype. The five sterility genes were located in five different linkage groups (Pitrat, 1991; Park et al., 2009). McCreight (1983) and Pitrat (1991) reported loose linkages between red stem (*r*) and the *ms-1* gene, and between yellow green leaves (*yg*) and the *ms-2* gene, respectively. Park et al. (2009) mapped the *ms-3* gene to the linkage group 9 of the linkage map Deltex x TGR-1551, which corresponds to the linkage group VII.

A *Macrocalyx* dominant gene, *Mca*, was reported to control the presence of large, leafy sepals in staminate and hermaphrodite flowers in the Japanese cultivar Makuwa (Ganesan and Sambandam, 1979). Two recessive genes were reported to modify the color of petals; *gp* (*green petals*) and *gyc* (*greenish yellow corolla*) control the presence of a green corolla with venation or the presence of a greenish yellow corolla, instead of the normal yellow corolla (Mockaitis and Kivilaan, 1965; Zink, 1986).

Rosa (1928) reported that tricarpellary ovary was monogenically inherited over pentacarpellary ovary found in Cassaba melons; the gene, named *p (pentamerous)* was mapped to the linkage group XII, closely linked to the major QTL for CMV resistance (Dogimont et al., 2000; Perin et al., 2002; Essafi et al., 2009). A single recessive gene, *n (nectarless)*, was reported to control the absence of nectar in all flowers in the mutant 40099 (Bohn, 1961).

Fruit Genes

Fruit shape was reported to be controlled by a single gene *O (Oval shape)*, dominant to round, and associated with *andromonoecious* gene *a* (Wall, 1967). As early as 1928, Rosa (1928) noted the association of elongate fruit with pistillate flowers (monoecious plants) and globular fruit with perfect flowers (andromonoecious plants) in segregating populations. More recently, several fruit shape QTL were mapped in several populations to at least five linkage groups; one of them co-localized with the *a* locus on the linkage group II (Perin et al., 2002; Monforte et al., 2004; Eduardo et al., 2007; Fernandez-Silva et al., 2010; Díaz et al., 2011). Spherical fruit shape was also reported to be controlled by a single gene, *sp* (*spherical fruit shape*), recessive to an obtuse fruit shape (Lumsden, 1914; Bains and Kang, 1963); this gene may be the same as the gene *O*.

A single dominant gene, *Ec (Empty cavity)*, was reported to control the presence of separated carpels at fruit maturity, leaving a cavity in PI 414723 fruit (*ec* in 'Védrantais') (Périn et al., 1999). The *Ec* gene was mapped to the linkage group III (Perin et al., 2002).

External fruit appearance. Rind color of melon fruit varieties includes white, yellow, orange, or green, and can be variegated. The white color of immature fruits was reported to be dominant to green immature fruits and controlled by a single gene, *Wi, White color of immature fruit* (Kubicki, 1962). The white color of mature fruits was, in contrast, reported to be controlled gene *w, white*, recessive to dark green fruit skin in a cross between Honeydew (*w*) and Smiths' Perfect cantaloupe (*W*, dark green) (Hughes, 1948). Melon rind color was shown to be based on different combinations of three major pigments, chlorophyll, carotenoids and naringerin-chalcone, a flavonoid pigment responsible for the yellow color of mature fruits in Yellow Canari melon type (Tadmor et al., 2010). Accumulation of naringerin-chalcone was reported to be inherited as a monogenic dominant trait in the cross 'Noy Amid' (yellow rind) x 'Tendral Verde Tardio' (dark green rind); accumulation of chlorophyll and carotenoids segregates jointly as a single dominant gene, independent to naringerin-chalcone accumulation (Tadmor et al., 2010). We propose to name *Nca* the gene, which regulates *naringerin-chalcone accumulation* (versus non-accumulation), and *Chl* and *Car* the two linked genes, which control *chlorophyll* and *carotenoid accumulation* in the rind of mature fruit, respectively. In addition, minor genes likely control quantitative variation of the accumulation of these pigments. A polygenic control of the external fruit color was reported in the cross 'Piel de Sapo' x PI 161375 (Whitaker and Davis, 1962; Monforte et al., 2004; Eduardo et al., 2007; Obando et al., 2008).

Vein tracts, formerly and incorrectly referred to as sutures, on the fruit rind was reported to be controlled by a single recessive gene *s, sutures* (Bains and Kang, 1963; Davis 1970). The same inheritance was found in two crosses: 'Védrantais' (*s-2*, presence of sutures) x PI 161375 (*S-2*, without sutures) and 'Védrantais' x PI 414723 (*S-2*). The *s-2* gene was mapped to the linkage group XI (Perin et al., 2002). Stripes on the rind was reported to have a monogenenic recessive inheritance (gene *st* for *striped epicarp*) by (Hagiwara and Kamimura, 1936). The presence of stripes on young fruits of 'Dulce' (before netting development) was also reported to be controlled by a single recessive gene, *st-2 (striped epicarp-2)*, in the cross Dulce (*st-2*) x PI 414723 (*St-2*, non-striped) (Danin-Poleg et al., 2002); the gene *st-2* was mapped to the linkage group XI. Further studies would be required to clarify the relationship between *st-2* and *s-2*, also located in the linkage group XI.

The ridge fruit surface was reported to be controlled by a single gene, *ri (ridge in C68)*, recessive to ridgeless (*Ri* in 'Pearl') (Takada et al., 1975). The speckled epidermis of the fruit is controlled by a single recessive gene, *spk* (*speckled fruit epidemis*) in PI 414723 (*Spk* in 'Védrantais') and was mapped to the linkage group VII (Perin et al., 2002). A single gene, *Mt* (*Mottled rind pattern*), was reported to control a mottled rind in 'Annamalai', dominant to uniform color *mt* in 'Makuwa' (Ganesan, 1988). The presence of dark spots (about 1 cm in diam.) on the rind (versus no spots) has a monogenic recessive inheritance in crosses Védrantais (*Mt-2*) x PI 161375 (*mt-2*) and Védrantais (*Mt-2*) x PI 414723 (*mt-2*), as the F_1 fruits have a uniform color rind (Périn et al., 1999); it was erroneously named *Mt-2* in the previous gene list. *mt-2* was mapped to the linkage group II (Perin et al., 2002).

A single dominant gene governing the development of net tissue, regardless of the degree of netting was reported in BIZ in a cross with smooth-skinned PI 414723 (Herman et al., 2008). We propose to name the gene *Rn* (*Rind netting*) instead of *N*. The gene was mapped to the linkage group II, closely linked to *fom1.2a* for Fusarium wilt resistance; additional minor loci likely affect the density of the net (Herman et al., 2008). Several QTL for the height and the width of the net in 'Deltex' were detected in a cross between netted 'Deltex' and net-free TGR-1551 (Park et al., 2009).

Melon fruit flesh color has been proposed to be controlled by two genes, *gf* for *green flesh* in Honeydew, recessive to orange flesh (*Gf* in Smiths' Perfect cantaloupe) (Hughes, 1948) and *wf* for *white flesh* (Iman et al., 1972). Genetic control of melon mesocarp color has, however, not been clearly elucidated and likely differs among market types. Clayberg (1992) confirmed that green and white mesocarps are recessive to orange and indicated that *gf* and *wf* interact epistatically. Mesocarp color (orange *vs.* green) segregated as a single recessive gene in recombinant inbred lines derived from orange flesh Védrantais x green flesh PI 161375 (Perin et al., 2002) and orange flesh AR 5 x green flesh Harukai N°3 (Fukino et al., 2008). The segregating gene, named *gf*, proposed to be renamed *wf*, mapped to the linkage group IX. In F_2 and doubled haploid lines derived from the cross between green mesocarp PI 161375 and white mesocarp Piel de Sapo T111, individuals with orange mesocarp were observed at a low frequency (Monforte et al., 2004); a single recessive gene segregated, if orange mesocarp phenotype was excluded and mapped to the linkage group VIII (formerly G1) (Monforte et al., 2004). Several QTL for fruit flesh color were described in near isogenic lines derived from the same cross (Eduardo et al., 2007; Obando et al., 2008). Recently, three QTL associated with color variation (white, green, orange) with putative epistatic interaction were identified in the cross between the white-fleshed Chinese line Q3-2-2 and orange-fleshed 'Top Mark' (Cuevas et al., 2009; Cuevas et

al., 2010). Five QTL associated with beta-carotene content, which is related to color intensity of the mesocarp, were identified in the cross between two orange-fleshed genotypes, USDA 846-1 and 'Top Mark' (Cuevas et al., 2008).

Sweet melon cultivars are characterized by high sucrose and low acid levels in mature fruit flesh. A single, incompletely recessive gene, *suc*, controlled accumulation of sucrose in the cross between the low sucrose Faqqous (var. *flexuosus*) and the high sucrose 'Noy Yizre'el' (Burger et al., 2002). Several QTL associated with total soluble solid content and sugar content have been described in several populations (Monforte et al., 2004; Sinclair et al., 2006; Park et al., 2009; Harel-Beja, 2010).

A dominant gene, *So* (*Sour*) was reported to control high acidity in melon fruit (Kubicki, 1962). A single dominant gene, *So-2* (*Sour-2*) for *sour taste* of the mature fruit, was also reported in PI 414723 (Périn et al., 1999; Burger et al., 2003). A single recessive gene, *pH*, was reported to control fruit flesh acidity in PI 414723. Low pH value in PI 4141723 was dominant to high pH value in 'Dulce'. The *pH* gene was mapped to the linkage group VIII (Danin-Poleg et al., 2002); it likely corresponds to *So-2*.

While ripe melon fruits usually do not have a bitter taste, young fruits are divided into two types: bitter and non-bitter. A single dominant gene, *Bif-1* (*Bitter fruit-1*, formely *Bif*), was reported to control the strong bitter taste of tender fruits in Indian wild melon (Parthasarathy and Sambandam, 1981). A monogenic dominant inheritance for the bitterness of young fruits was confirmed in wild melons from Africa and China (Ma et al., 1997). The cross of non-bitter melon lines (var. *conomon* and var.*makuwa*) with var. *inodorus* and var. *cantalupensis*) yielded, however, bitter young melons, which suggests complementary gene action of two independent genes, *Bif-2* and *Bif-3* (*Bif-2_ Bif-3_* are bitter; *bif-2bif-2 Bif-3_* and *Bif-2_ bif-3bif-3* are non-bitter) (Ma et al., 1997). One of them may be the same as *Bif-1*. The relationship with the gene *Bi* controlling seedling bitterness (Lee and Janick, 1978) is unknown.

While the single dominant gene *Mealy, Me*, was reported to control mealy flesh texture by Ganesan (1988) in an accession named *C. callosus* crossed with a crisp-fleshed 'Makuwa', a monogenic recessive inheritance was found for the mealy flesh texture in the var. *momordica* accession PI 414723 (*me-2*) crossed by 'Védrantais' (*Me-2*) (Périn et al., 1999) (included erroneously as *Me-2* in the previous gene list, it is now included as *me-2*). A monogenic recessive inheritance was reported for the juicy character of melon fruit flesh; the gene was named *juicy flesh*, symbolized *jf* (Chadha et al., 1972). A single gene was reported to control the musky flavor of *C. melo callosus (Mu, Musky)*, dominant to the mild flavor in 'Makuwa' or 'Annamalai' (*mu*) (Ganesan, 1988).

Fruit abscission at maturity was reported to be controlled by two independent loci in two independent studies. In absence of allelism tests, the genes were named abscission layer *Al-1* and *Al-2* in C68, *al-1* and *al-2* in 'Pearl' (Takada et al., 1975), and *Al-3* and *Al-4* in the climacteric Charentais type 'Védrantais' (Perin et al., 2002). *Al-3* and *Al-4* were mapped to the linkage groups VIII and IX in a recombinant inbred population derived from a cross between 'Védrantais' and the non-climacteric PI 161375 (Perin et al., 2002). A single dominant gene, *Al-5*, was reported to control fruit abscission layer formation in the climacteric western shipper type 'TAM Uvalde' in the cross with the non-climacteric Casaba type 'TAM Yellow Canary' (Zheng et al., 2002).

Organogenic competence varies among melon genotypes. *In vitro* shoot regeneration capacity was reported to be controlled by two independent genes, partially dominant, *Org-1* and *Org-2* (*Organogenic* response) (Molina and Nuez, 1996). A single dominant gene, *Org-3*, was reported to control the high regeneration competence in the line BU-21/3, in crosses with the low regeneration competent lines 'PMR 45' and 'Ananas Yokneam' (Galperin et al., 2003).

Table 1. Reported host plant resistance and morphological genes of melon, including genes symbol, synonyms, descriptions, and linkage groups.[z]

Gene symbol Prefered	Synonym	Character	LG[y]	References
a	*M*	*andromonoecious.* **Mostly staminate, fewer perfect flowers; on *A_* plants, pistillate flowers have no stamens; epistatic to *g*.**	4, II	(Rosa, 1928 ; Poole and Grimball, 1939; Wall, 1967)
ab	-	*abrachiate.* Lacking lateral branches. Interacts with *a* and *g*, e.g., *abab aa G_* plants produce only staminate flowers.		(Foster and Bond, 1967)
Ac	-	*Alternaria cucumerina* **resistance, in MR-1.**		(Thomas et al., 1990)
Af	-	*Aulacophora foveicollis* resistance. Resistance to the red pumpkin beetle.		(Vashistha and Choudhury, 1974)
Ag	-	*Aphis gossypii* **tolerance. Freedom of leaf curling following aphid infestation; in PI 414723.**		(Bohn et al., 1973)
Ala	-	*Acute leaf apex.* Dominant over obtuse apex, linked with *Lobed* leaf, *Ala* in Maine Rock, *ala* in PV Green.		(Ganesan and Sambandam, 1985)
alb	-	*albino.* **White cotyledons, lethal mutant; in Trystorp.**		(Besombes et al., 1999)
Al-1	*Al₁*	*Abscission layer-1.* One of two dominant genes for abscission layer formation, *Al-1Al-2* in C68, *al-1al-2* in Pearl. See *Al-2*.		(Takada et al., 1975)
Al-2	*Al₁₂*	*Abscission layer-2.* One of two dominant genes for abscission layer formation. See *Al-1*.		(Takada et al., 1975)
Al-3		*Abscission layer-3.* **One dominant gene for abscission layer formation in PI 161375. Relationship with *Al-1* or *Al-2* is unknown**	VIII	(Perin et al., 2002)
Al-4		*Abscission layer-4.* **One dominant gene for abscission layer formation in PI 161375. Relationship with *Al-1* or *Al-2* is unknown**	IX	(Perin et al., 2002)
Al-5	-	*Abscission layer-5.* **One dominant gene for abscission layer formation; full-slip in TAM Uvalde.**		(Zheng et al., 2002)
bd	-	*brittle dwarf.* Rosette growth with thick leaf. Male fertile, female sterile; in TAM-Perlita 45.		(Cox, 1985)
Bi	-	*Bitter.* **Bitter seedling. Common in honeydew or in Charentais type while most American cantaloupes are *bi*.**		(Lee and Janick, 1978)
Bif-1	*Bif*	*Bitter fruit-1.* Bitterness of tender fruit in wild melon. Relation with *Bi* is unknown.		(Parthasarathy and Sambandam, 1981)
Bif-2	-	*Bitter fruit-2.* One of two complementary independent genes for bitter taste in young fruit: *Bif-2_ Bif-3_* are bitter. Relationships with *Bi* and *Bif-1* are unknown.		(Ma et al., 1997)
Bif-3	-	*Bitter fruit-3.* One of two complementary independent genes for bitter taste in young fruit: *Bif-2_ Bif-3_* are bitter. Relationships with *Bi* and *Bif-1* are unknown.		(Ma et al., 1997)
cab-1	-	*cucurbit aphid borne* **yellows virus resistance-1. One of two complementary independent genes for resistance to this polerovirus: *cab-1cab-1 cab-2cab-2* plants are resistant; in PI 124112.**		(Dogimont et al., 1997)
cab-2	-	*cucurbit aphid borne* **yellows virus resistance-2. One of two complementary independent genes for resistance to this polerovirus: *cab-1cab-1 cab-2cab-2* plants are resistant; in PI 124112.**		(Dogimont et al., 1997)
Car		*Carotenoids accumulation* **in the rind of mature fruit. A single dominant gene for the accumulation of these pigments in the rind of mature fruit versus non accumulation; *Car* in Tendral Verde Tardio, *car* in Noy Amid (Canary type); linked to *Chl*.**		(Tadmor et al., 2010)

| Gene symbol | | Character | LG[y] | References |
Prefered	Synonym			
cb	cb[l]	*cucumber beetle* resistance. Interacts with *Bi*, the nonbitter *bibi cbcb* being the more resistant; in C922-174-B.		(Nugent et al., 1984)
cf	-	**cochleare folium. Spoon-shaped leaf with upward curling of the leaf margins; spontaneous mutant in Galia.**		(Lecouviour et al., 1995)
cgmmv-1		*cucumber green mottle mosaic virus resistance-1*. One of two complementary genes for resistance to this tobamovirus in Chang Bougi.		(Sugiyama et al., 2007)
cgmmv-2		*cucumber green mottle mosaic virus resistance-2*. One of two complementary genes for resistance to this tobamovirus in Chang Bougi.		(Sugiyama et al., 2007)
Chl		**Chlorophyll accumulation in the rind of mature fruit. A single dominant gene for chlorophyll accumulation in the rind of mature fruit versus non-accumulation; Chl in Tendral Verde Tardio, chl in Noy Amid, (Canary type), linked to Car.**		(Tadmor et al., 2010)
cl	-	*curled leaf.* Elongated leaves that curl upward and inward. Usually male and female sterile.		(Cox, 1985)
Creb-2		*Cucumber mosaic virus resistance.* A single dominant gene for resistance to this cucumovirus in Yamatouri		(Daryono et al., 2010)
culcrv		**cucurbit leaf crumple virus resistance. A single recessive gene for resistance to this geminivirus transmitted by whitefly in PI 313970.**		(McCreight et al., 2008)
Cys	-	**Cucurbit yellow stunting disorder virus resistance. One dominant gene for resistance to this crinivirus in TGR-1551.**		(Lopez-Sese and Gomez-Guillamon, 2000)
dc-1	-	*Dacus cucurbitae-1* resistance. One of two complementary recessive genes for resistance to the melon fruitfly. See *dc-2*.		(Sambandam and Chelliah, 1972)
dc-2	-	*Dacus cucurbitae-2* resistance. One of two complementary recessive genes for resistance to the melon fruitfly. See *dc-1*.		(Sambandam and Chelliah, 1972)
dl	-	**dissected leaf. Highly indented leaves in URSS 4.**		(Dyutin, 1967)
dl[v]	cl	**dissected leaf Velich.** First described as *cut leaf* in Cantaloup de Bellegarde. **Allelic to dl.**		(Velich and Fulop, 1970)
dl-2	-	*dissected leaf-2*. First described as « hojas hendidas ».		(Esquinas Alcazar, 1975)
dlet	dl	*delayed lethal*. Reduced growth, necrotic lesions on leaves and premature death.		(Zink, 1990)
Ec	-	**Empty cavity. Carpels are separated at fruit maturity leaving a cavity; Ec in PI 414723, ec in Védrantais.**	III	(Périn et al., 1999)
ech	-	**exaggerated curvature of the hook. Triple response of seedlings germinating in darkness in presence of ethylene; ech in PI 161375, Ech in Védrantais.**	I	(Perin et al., 2002)
f	-	**flava. Chlorophyl deficient mutant. Growth rate reduced in K 2005.**	8	(Pitrat et al., 1986)
fas	-	**fasciated stem, in Vilmorin 104.**		(Gabillard and Pitrat, 1988)
fe	-	**fe (iron) inefficient mutant. Chlorotic leaves with green veins that turn green when adding iron in the nutrient solution.**		(Nugent and Bhella, 1988; Jolley et al., 1991)
Fn	-	**Flaccida necrosis. Semi-dominant gene for wilting and necrosis reactions to F pathotype of *Zucchini yellow mosaic virus*; Fn in Doublon, fn in Védrantais).**	2, V	(Risser et al., 1981)
Fom-1	Fom[1]	**Fusarium oxysporum melonis resistance. Resistance to races 0 and 2 and susceptibility to races 1 and 1.2 of Fusarium wilt; Fom-1 in Doublon, fom-1 in Charentais T.**	5, IX	(Risser, 1973)

Gene symbol		Character	LG[y]	References
Prefered	**Synonym**			
Fom-2	*Fom1.2*	***Fusarium oxysporum melonis* resistance. Resistance to races 0 and 1 and susceptibility to races 2 and 1.2 of Fusarium wilt; *Fom-2* in CM 17187, *fom-2* in Charentais T.**	6, XI	(Risser, 1973)
Fom-3	-	***Fusarium oxysporum melonis* resistance. Same phenotype as *Fom-1* but segregates independently from *Fom-1*; *Fom-3* in Perlita FR, *fom-3* in Charentais T.**		(Zink and Gubler, 1985)
fom1.2a		*Fusarium oxysporum melonis* resistance. Resistance to race 1.2 of Fusarium wilt; *fom1.2a* in BIZ, *Fom1.2a* in PI 414723.	II	(Herman et al., 2008)
fom-4		*Fusarium oxysporum melonis* resistance. Resistance to race 0 and 2 of Fusarium wilt, *fom-4* in Tortuga, likely associated with *Fom-1*.		(Oumouloud et al., 2010)
g	-	*gynoecious*. Controls the presence of one (*g*) or two (*G*) types of flowers on one plant. Epistatic to *a*: *A_ G_* monoecious; *A_ gg* gynoecious; *aa G_* andromonoecious; *aa gg* hermaphrodite.		(Poole and Grimball, 1939)
gf	-	*green flesh* color. Recessive to salmon, *gf* in honeydew, *Gf* in Smiths' Perfect cantaloupe.	VIII	(Hughes, 1948)
gl	-	*glabrous*. Trichomes lacking in Arizona glA.	3	(Foster, 1963)
gp	-	*green petals*. Corolla leaf like in color and venation.		(Mockaitis and Kivilaan, 1965)
Gs	-	*Gelatinous sheath* around the seeds. Dominant to absence of gelatinous sheath.		(Ganesan, 1988)
Gsb-1	*Mc*	***Gummy stem blight* resistance-1. High degree of resistance to *Didymella bryoniae* (= *Mycosphaerella citrullina*) in PI 140471.**		(Prasad and Norton, 1967; Frantz and Jahn, 2004)
Gsb-2	*Mc-3*	*Gummy stem blight* resistance-2. High level of resistance to *Didymella bryoniae* (= *Mycosphaerella citrullina*) in PI 157082), independent from *Gsb-1, Gsb-3, Gsb-4* and *gsb-5*.		(Zuniga et al., 1999; Frantz and Jahn, 2004)
Gsb-3	*Mc-4-*	***Gummy stem blight* resistance-3. High level of resistance to *Didymella bryoniae* (= *Mycosphaerella citrullina*) in PI 511890, independent from *Gsb-1, Gsb-2, Gsb-4* and *gsb-5*.**		(Zuniga et al., 1999; Frantz and Jahn, 2004)
Gsb-4	-	***Gummy stem blight* resistance-4. High level of resistance to *Didymella bryoniae* (= *Mycosphaerella citrullina*) in PI 482398, independent from *Gsb-1, Gsb-2, Gsb-3* and *gsb-5*.**		(Frantz and Jahn, 2004)
gsb-5	-	***gummy stem blight* resistance-5. High level of resistance to *Didymella bryoniae* (= *Mycosphaerella citrullina*) in PI 482399, independent from *Gsb-1, Gsb-2, Gsb-3* and *Gsb-4*.**		(Frantz and Jahn, 2004)
Gsb-6	*Mc[i], Mc-2*	*Mycosphaerella citrullina* resistance-2. Moderate degree of resistance to gummy stem blight in C-1 and C-8.		(Prasad and Norton, 1967)
gyc	-	***greenish yellow corolla.***		(Zink, 1986)
gy	*n, M*	***gynomonoecious*. Interacts with *a* and *g* to produce stable gynoecious plants (*A_ g g gy gy*) in WI 998.**		(Kenigsbuch and Cohen, 1987, 1990)
h	-	***halo* cotyledons. Yellow halo on the cotyledons, later turning green.**	4, II	(Nugent and Hoffman, 1974)
Imy	-	*Interveinal mottling and yellowing* resistance. Resistance to a complex of viruses, in PI 378062.		(Hassan et al., 1998)
jf	-	*juicy flesh*. Segregates discretely in a monogenic ratio in segregating generations.		(Chadha et al., 1972)
L	-	*Lobed* leaf. Dominant on non lobed, linked with *Acute leaf apex L* in Maine Rock, *l* in P.V. Green.		(Ganesan and Sambandam, 1985)

Gene symbol				
Prefered	Synonym	Character	LG[y]	References
Liy	-	*Lettuce infectious yellows* virus resistance. One dominant gene for resistance to this crinivirus in PI 313970.		(McCreight, 2000)
lmi	-	*long mainstem internode*. Affects internode length of the main stem but not of the lateral ones in 48764.	8	(McCreight, 1983)
Lt	-	*Liriomyza trifolii* (leafminer) resistance in Nantais Oblong.		(Dogimont et al., 1999)
M-Pc-5	-	*Modifier of Pc-5*. Gene *Pc-5* for downy mildew resistance is dominant in presence of *M-Pc-5*, recessive in the absence of *M-Pc-5*.		(Angelov and Krasteva, 2000)
Mca	-	*Macrocalyx*. Large, leaf like structure of the sepals in staminate and hermaphrodite flowers; *Mca* in makuwa, *mca* in Annamalai.		(Ganesan and Sambandam, 1979)
Me	-	*Mealy* flesh texture. Dominant to crisp flesh; *Me* in *C. callosus*, *me* in makuwa.		(Ganesan, 1988)
me-2	-	*mealy* flesh texture-2 in PI 414723.		(Périn et al., 1999)
Mnr-1	*Mnr1*	*Melon necrotic resistance 1*. One of two dominant genes for resistance to *Melon necrotic spot virus* (MNSV) located at 19 cM from *nsv*; *Mnr-1* in Doublon, *mnr-1* in ANC-42.	XII	(Mallor Gimenez et al., 2003)
Mnr-2	*Mnr2*	*Melon necrotic resistance 2*. One of two dominant genes for resistance to *Melon necrotic spot virus* (MNSV) independent from *Mnr-1*; *Mnr-2* in Doublon, *mnr-2* in ANC-42.		(Mallor Gimenez et al., 2003)
ms-1	*ms[1]*	*male sterile-1*. Indehiscent anthers with empty pollen walls in tetrad stage.	3	(Bohn and Whitaker, 1949)
ms-2	*ms[2]*	*male sterile-2*. Anthers indehiscent, containing mostly empty pollen walls, growth rate reduced.	6, XI	(Bohn and Principe, 1964)
ms-3	*ms-L*	*male sterile-3*. Waxy and translucent indehiscent anthers, containing two types of empty pollen sacs.	12, VII	(McCreight and Elmstrom, 1984)
ms-4	-	*male sterile-4*. Small indehiscent anthers. First male flowers abort at bud stage in Bulgaria 7.	9	(Lozanov, 1983)
ms-5	-	*male sterile-5*. Small indehiscent anthers. Empty pollen in Jivaro, Fox.	13	(Lecouviour et al., 1990)
Mt	-	*Mottled* rind pattern. Dominant to uniform color. Epistatic with *Y* (not expressed in *Y_*) and *st* (*Mt_ st st* and *Mt_ St_* mottled; *mt mt st st* striped, *mt mt St_* uniform); *Mt* in Annamalai, *mt* in Makuwa.		(Ganesan, 1988)
mt-2	-	*mottled* rind pattern in PI 161375.	II	(Périn et al., 1999)
Mu	-	*Musky* flavor (olfactory). Dominant on mild flavor; *Mu* in *C. melo callosus*, *mu* in Makuwa or Annamalai.		(Ganesan, 1988)
Mvd	-	*Melon vine decline* resistance in Pat 81. Semi-dominant gene for partial resistance to *Acremonium cucurbitacearum* and *Monosporascus cannonballus*,.		(Iglesias et al., 2000)
My	-	*Melon yellows* virus resistance. Semi-dominant gene for partial resistance to this crinivirus, in Nagata Kin Makuwa.		(Esteva and Nuez, 1992; Nuez et al., 1999)
n	-	*nectarless*. Nectaries lacking in all flowers of 40099.		(Bohn, 1961)
Nca		*naringerin-chalcone accumulation* in the rind of mature fruit. A single dominant gene for the accumulation of this flavonoid pigment in the rind of mature fruit versus non-accumulation; *Nca* in Noy Amid, Canary type, *nca* in Tendral Verde Tardio.		(Tadmor et al., 2010)
Nm	-	*Necrosis* with *Morocco* strains of *Watermelon mosaic virus*, a potyvirus; *Nm* in Védrantais, *nm* in Ouzbèque.		(Quiot-Douine et al., 1988)

Gene symbol				
Prefered	Synonym	Character	LG[y]	References
nsv	-	Melon *necrotic spot virus* resistance. A single recessive gene for resistance to this carmovirus in Gulfstream, Planters Jumbo.	7, XII	(Coudriet et al., 1981)
O	-	*Oval* fruit shape. Dominant to round, associated with *a*.		(Wall, 1967)
Org-1	-	*Organogenic* response for *in vitro* shoot regeneration. Partially dominant. Interacts with an additive model with *Org-2*.		(Molina and Nuez, 1996)
Org-2	-	*Organogenic* response for *in vitro* shoot regeneration. Partially dominant. Interacts with an additive model with *Org-1*.		(Molina and Nuez, 1996)
Org-3	-	*Organogenic* response for *in vitro* regeneration. Dominant allele for high response in BU-12/3, recessive allele in PMR 45 or Ananas Yokneam. Probably different from *Org-1* and *Org-2*.		(Galperin et al., 2003)
p	-	*pentamerous*. Five carpels and stamens; recessive to trimerous; in Casaba.	XII	(Rosa, 1928)
Pa	-	*Pale* green foliage. *PaPa* plants are white (lethal); *Papa* are yellow; in 30567.	3	(McCreight and Bohn, 1979)
Pc-1	-	*Pseudoperonospora cubensis* resistance. One of two complementary incompletely dominant genes for downy mildew resistance in PI 124111. See *Pc-2*.		(Cohen et al., 1985; Thomas et al., 1988)
Pc-2	-	*Pseudoperonospora cubensis* resistance. One of two complementary incompletely dominant genes for downy mildew resistance in PI 124111). See *Pc-1*.		(Cohen et al., 1985; Thomas et al., 1988)
Pc-3	-	*Pseudoperonospora cubensis* resistance. Partial resistance to downy mildew in PI 414723.		(Epinat and Pitrat, 1989)
Pc-4	-	*Pseudoperonospora cubensis* resistance. One of two complementary genes for downy mildew resistance in PI 124112. Interacts with *Pc-1* or *Pc-2*.		(Kenigsbuch and Cohen, 1992)
Pc-5	-	*Pseudoperonospora cubensis* resistance. One gene in Line 5-4-2-1 which interacts with *M-Pc-5* in the susceptible line K15-6; *Pc-5* is dominant in presence of *M-Pc-5*, recessive in the absence of *M-Pc-5*.		(Angelov and Krasteva, 2000)
pH	-	*pH* (acidity) of the mature fruit flesh. Low pH value in PI 414723 dominant to high pH value in Dulce.	VIII	(Danin-Polog et al., 2002)
pin	-	*pine-seed* shape in PI 161375.	III	(Perin et al., 2002)
Pm-1	*Pm[1]* *Pm-A ?*	*Powdery mildew* resistance-1. Resistance to race 1 of *Podosphaera xanthi* in PMR 45.		(Jagger et al., 1938)
Pm-2	*Pm[2]* *Pm-C ?*	*Powdery mildew* resistance-2. Interacts with *Pm-1*; Resistance to race 2 of *Podosphaera xanthii* in PMR 5 with *Pm-1*.		(Bohn and Whitaker, 1964)
Pm-3	*Pm[3]*	*Powdery mildew* resistance-3. Resistance to race 1 of *Podosphaera xanthii* in PI 124111.	7	(Harwood and Markarian, 1968, 1968)
Pm-4	*Pm[4]*	*Powdery mildew* resistance-4. Resistance to *Podosphaera xanthii* in PI 124112.		(Harwood and Markarian, 1968, 1968)
Pm-5	*Pm[5]*	*Powdery mildew* resistance-5. Resistance to *Podosphaera xanthii* in PI 124112.		(Harwood and Markarian, 1968, 1968)
Pm-6	-	*Powdery mildew* resistance-6. Resistance to *Podosphaera xanthii* race 2 in PI 124111.		(Kenigsbuch and Cohen, 1989)
Pm-7	-	*Powdery mildew* resistance-7. Resistance to *Podosphaera xanthii* race 1 in PI 414723.		(Anagnostou et al., 2000)

Gene symbol				
Prefered	**Synonym**	**Character**	**LG**[y]	**References**
Pm-8		*Powdery mildew* resistance-8.. Resistance to *Podosphaera xanthii* race pxCh1 in PI 134198.		(Liu et al., 2010)
Pm-E	-	*Powdery mildew* resistance-E. Interacts with *Pm-C* in PMR 5 for *Golovinomyces cichoracearum* resistance.		(Epinat et al., 1993)
Pm-F	-	*Powdery mildew* resistance-F. Interacts with *Pm-G* in PI 124112 for *Golovinomyces cichoracearum* resistance.		(Epinat et al., 1993)
Pm-G	-	*Powdery mildew* resistance-G. Interacts with *Pm-F* in PI 124112 for *Golovinomyces cichoracearum* resistance.		(Epinat et al., 1993)
Pm-H	-	*Powdery mildew* resistance-H. Resistance to *Golovinomyces cichoracearum* and susceptibility to *Podosphaera xanthii* in Nantais oblong.		(Epinat et al., 1993)
Pm-R	-	*Powdery mildew* resistance-R. Resistance to *Podosphaera xanthii* races 1, 2, and 5 in TGR-1551.	V	(Yuste-Lisbona et al., 2011)
pm-S		*powdery midew* resistance-S. Resistance to *Podosphaera xanthii* race S in PI 313970. Recessive to susceptibility in Top Mark.		(McCreight and Coffey, 2011)
Pm-w	*Pm-B ?*	*Powdery mildew* resistance-*w*. Resistance to *Podosphaera xanthii race 2* in WMR 29.	2, V	(Pitrat, 1991)
Pm-x	-	*Powdery mildew* resistance-*x*. Resistance to *Podosphaera xanthii* in PI 414723.	4, II	(Pitrat, 1991)
Pm-y	-	*Powdery mildew* resistance-*y*. Resistance to *Podosphaera xanthii* in VA 435.	7, XII	(Pitrat, 1991)
Pm-z		*Powdery mildew* resistance-*z*. Resistance to *Podosphaera xanthii* races 1 and 2US in PI 313970.		(McCreight, 2003)
PmV.1	-	*Powdery mildew* resistance V.1. Resistance to *Podosphaera xanthii* races 1, 2 and 3 in PI 124112.	V	(Perchepied et al., 2005)
PmXII.1		*Powdery mildew* resistance XII.I. Resistance to *Podosphaera xanthii* races 1, 2 and 5 and to *Golovinomyces cichoracearum* race 1 in PI 124112	XII	(Perchepied et al., 2005)
Prv[1]	*Wmv*	*Papaya Ringspot virus* resistance[1]. Resistance to W strain of this potyvirus (formerly *Watermelon mosaic virus 1*) in PI 180280, and B66-5, WMR 29, which were derived from PI 180280. Dominant to *Prv*[2].	5, IX	(Webb, 1979; Pitrat and Lecoq, 1983)
Prv[2]	-	*Papaya Ringspot virus* resistance[2]. Allele at the same locus as *Prv*[1] but different reaction with some strains of the virus; in 72-025, which was derived from PI 180283. Recessive to *Prv*[1].	5, IX	{Kaan, 1973); Pitrat, 1983}
Prv-2	-	*Papaya Ringspot virus* resistance-2. Relationship with *Prv* is unknown; in PI 124112		(McCreight and Fashing-Burdette, 1996)
r	-	*red* stem. Red pigment under epidermis of stems, especially at nodes, and reddish or tan seed color; in PI 157083.	3	(Bohn, 1968; McCreight and Bohn, 1979)
ri	-	*ridge*. Ridged fruit surface, recessive to ridgeless. (*ri* in C68, *Ri* in Pearl).		(Takada et al., 1975)
Rn	*N*	*Rind netting*. Netted fruit surface, regardless of the degree of netting; *Rn* in B12 dominant to smooth, non-netted rind, *rn* in PI 414723.		(Herman et al., 2008)
s	-	*sutures*. Presence of vein tracts on the fruit (« sutures »); recessive to ribless.		(Bains and Kang, 1963)
s-2	-	*sutures-2* on the fruit rind of PI 161375. Relationship with *s* is unknown.	XI	(Périn et al., 1999)
Sfl	*S*	*Subtended floral leaf*. The floral leaf bearing the hermaphrodite flowers is sessile, small and encloses the flower; *Sfl* in makuwa, *sfl* in Annamalai.		(Ganesan and Sambandam, 1979)

Gene symbol				
Prefered	Synonym	Character	LG[y]	References
si-1	*b*	*short internode-1*. Extremely compact plant habit (bush type) in UC Topmark Bush.	1	(Denna, 1962)
si-2	-	*short internode-2*. Short internodes from 'birdnest' melon in Persia 202.		(Paris et al., 1984)
si-3	-	*short internode-3*. Short internodes in Maindwarf.		(Knavel, 1990)
slb	*sb*	*short lateral branching*. Reduction of the elongation of the lateral branches, in LB-1		(Ohara et al., 2001)
So	-	*Sour* taste. Dominant to sweet.		(Kubicki, 1962)
So-2	-	*Sour* taste-2. Relationship with *So* is unknown, in PI 414723.		(Périn et al., 1999)
sp	-	*spherical* fruit shape. Recessive to obtuse; dominance incomplete.		(Lumsden, 1914; Bains and Kang, 1963)
spk	-	*speckled* fruit epidermis; *spk* in PI 161375 and PI 414723, *Spk* in Védrantais.	VII	(Perin et al., 2002)
sqmv		*squash mosaic virus* resistance. A single recessive gene for resistance to this Comovirus in China 51.		(Provvidenti, 1998)
st	-	*striped* epicarp. Recessive to non-striped.		(Hagiwara and Kamimura, 1936)
st-2	*st*	*striped epicarp-2*. Present in Dulce, recessive to non-striped in PI 414723. Relationship with *st* is unknown.	XI	(Danin-Poleg et al., 2002)
suc		*sucrose* accumulation. Low sucrose level in Faqqous (*suc*), high sucrose in Noy Yizre'el (*Suc*). Incomplete recessivity.		(Burger et al., 2002)
v	-	*virescent*. Pale cream cotyledons and hypocotyls, and yellow green foliage, mainly young leaves.		(Hoffman and Nugent, 1973)
v-2	-	*virescent-2*.		(Dyutin, 1979)
v-3	-	*virescent-3*. White cotyledons which turn green, light green young leaves which are normal when they are older.		(Pitrat et al., 1995)
Vat	-	*Virus aphid transmission* resistance. Resistance to several viruses when transmitted by *Aphis gossypii*, in PI 161375.	2, V	(Pitrat and Lecoq, 1980)
w	-	*white* color of mature fruit. Recessive to dark green fruit skin; *w* in honeydew, *W* in Smiths' Perfect cantaloupe.		(Hughes, 1948)
wf	-	*white flesh*. Recessive to salmon. *Wf* epistatic to *Gf_*.	IX	(Iman et al., 1972; Clayberg, 1992)
Wi	-	White color of *immature* fruit. Dominant to green.		(Kubicki, 1962)
Wmr	-	*Watermelon mosaic virus* (formerly *Watermelon mosaic virus 2*) resistance. A single dominant gene, in PI 414723	II	(Gilbert et al., 1994)
wmr-2	-	*Watermelon mosaic virus* (formerly *Watermelon mosaic virus 2*) resistance. A single recessive gene, in TGR-1551.		(Diaz-Pendon et al., 2005)
Wt	-	*White testa*. Dominant to yellow or tan seed coat color.		(Hagiwara and Kamimura, 1936)
Wt-2	-	*White testa-2*. Relationship with *Wt* unknown, in PI 414723.	IV	(Périn et al., 1999)
Y	-	*Yellow* epicarp. Dominant to white fruit skin.		(Hagiwara and Kamimura, 1936)

Gene symbol		Character	LG[y]	References
Prefered	Synonym			
yg	-	*yellow green* leaves. Reduced chlorophyll content.	6, XI	(Whitaker, 1952)
yg[w]	*lg*	*yellow green Weslaco*. First described as *light green* in a cross Dulce x TAM-Uvalde. Allelic to *yg*.		(Cox and Harding, 1986)
yv	-	*yellow virescence*. Pale cotyledons; yellow green young leaves and tendrils; bright and yellow petals and yellow stigma; etiolated; older leaves becoming green.	1	(Zink, 1977)
yv-2	*yv-X*	*yellow virescence-2*. Young leaves yellow green, old leaves normal green	5, IX	(Pitrat, 1991)
Zym	**Zym-1**	*Zucchini Yellow Mosaic* virus resistance. Resistance to pathotype 0 of this potyvirus in PI 414723.	4, II	(Pitrat and Lecoq, 1984)
Zym-2	-	*Zucchini Yellow Mosaic* potyvirus resistance. One of three complementary genes (see *Zym* and *Zym-3*) for resistance to this potyvirus in PI 414723.		(DaninPoleg et al., 1997)
Zym-3	-	*Zucchini Yellow Mosaic* potyvirus resistance. One of three complementary genes (see *Zym* and *Zym-2*) for resistance to this potyvirus in PI 414723.		(DaninPoleg et al., 1997)

Maternally inherited genes.

cyt-Yt	-	*cytoplasmic yellow tip*. Chlorophyll deficient mutant with yellow young leaves, turning green when becoming older. Maternally inherited.		(Ray and McCreight, 1996)

[z] Genes maintained by the curators or very common in collections (like *andromonoecious* or *white testa*) are **emboldened**. Genes that have been apparently lost or not maintained by curators, or have uncertain descriptions are normal in weight.

[y]Linkage groups to which the genes belong are indicated as Arabic numbers, according to Pitrat (1991) and Roman numbers according to Périn et al. (2002).

Table 2. Reported melon isozyme genes, including gene symbol, synonym, description, location on the melon genome and reference.

Gene symbol		Gene description and type lines	LG[z]	References
Prefered	Synonym			
Aco-1	*Ac*	*Aconitase-1.* Isozyme variant with two alleles, each regulating one band, in PI 218071, PI 224769.	A	(Staub et al., 1998)
Acp-1	*APS-11, Ap-11*	*Acid phosphatase-1.* Isozyme variant with two codominant alleles, each regulating one band. The heterozygote has two bands.		(Esquinas Alcazar, 1981)
Acp-2	*Acp-1*	*Acid phosphatase-2.* Isozyme variant with two alleles, each regulating one band, in PI 194057, PI 224786. Relationship with *Acp-1* is unknown.		(Staub et al., 1998)
Acp-4	-	*Acid phosphatase-4.* Isozyme variant with two alleles, each regulating one band, in PI 183256, PI 224786. Relationship with *Acp-1* unknown, different from *Acp-2*.		(Staub et al., 1998)
Ak-4	-	*Adenylate kinase.* Isozyme variant with two alleles, each regulating one band, in PI 169334.		(Staub et al., 1998)
Fdp-1	-	*Fructose diphosphate-1.* Isozyme variant with two alleles, each regulating one band, in PI 218071, PI 224688.		(Staub et al., 1998)
Fdp-2	-	*Fructose diphosphate-2.* Isozyme variant with two alleles, each regulating one band, in PI 204691, PI 183256.		(Staub et al., 1998)
Gpi	-	*Glucosephosphate isomerase.* Isozyme variant with two alleles, each regulating one band, in PI 179680.		(Staub et al., 1998)
Idh	-	*Isocitrate dehydrogenase.* Isozyme variant with two alleles, each regulating one band, in PI 218070, PI 224688.	A	(Staub et al., 1998)
Mdh-2	-	*Malate dehydrogenase-2.* Isozyme variant with two alleles, each regulating one band, in PI 224688, PI 224769.	B	(Staub et al., 1998)
Mdh-4	-	*Malate dehydrogenase-4.* Isozyme variant with two alleles, each regulating one band, in PI 218070, PI 179923.	B	(Staub et al., 1998)
Mdh-5	-	*Malate dehydrogenase-5.* Isozyme variant with two alleles, each regulating one band, in PI 179923, PI 180283.	B	(Staub et al., 1998)
Mdh-6	-	*Malate dehydrogenase-6.* Isozyme variant with two alleles, each regulating one band, in PI 179923, PI 180283.	B	(Staub et al., 1998)
Mpi-1	-	*Mannosephosphate isomerase-1.* Isozyme variant with two alleles, each regulating one band, in PI 183257, PI 204691.	A	(Staub et al., 1998)
Mpi-2	-	*Mannosephosphate isomerase-2.* Isozyme variant with two alleles, each regulating one band, in PI 183257, PI 204691.	A	(Staub et al., 1998)
Pep-gl	-	*Peptidase with glycyl-leucine.* Isozyme variant with two alleles, each regulating one band, in PI 218070.	B	(Staub et al., 1998)
Pep-la	-	*Peptidase with leucyl-alanine.* Isozyme variant with two alleles, each regulating one band, in PI 183256.		(Staub et al., 1998)
Pep-pap	-	*Peptidase with phenylalanyl-proline.* Isozyme variant with two alleles, each regulating one band, in PI 183256.		(Staub et al., 1998)
Pgd-1	*6-PGDH-21 Pgd-21*	*Phosphoglucodehydrogenase-1.* Isozyme variant with two alleles, each regulating one band.The heterozygote has one intermediate band.		(Esquinas Alcazar, 1981)
6-Pgd-2	-	*6-Phosphogluconate dehydrogenase.* Isozyme variant with two alleles, each regulating one band, in PI 161375, Védrantais. Relationship with *Pgd-1* is unknown.	IX	(Baudracco-Arnas and Pitrat, 1996)

| Gene symbol | | Gene description and type lines | LG[z] | References |
Prefered	Synonym			
Pgd-3	*Pgd*	*6-Phosphogluconate dehydrogenase.* Isozyme variant with two alleles, each regulating one band, in PI 218070. Relationship with *Pgd-1* and *6-Pgd-2* is unknown.	A	(Staub et al., 1998)
Pgi-1	*PGI-11*	*Phosphoglucoisomerase-1.* Isozyme variant with two alleles, each regulating two bands. The heterozygote has three bands.		(Esquinas Alcazar, 1981)
Pgi-2	*PGI-21*	*Phosphoglucoisomerase-2.* Isozyme variant with two alleles, each regulating two bands. The heterozygote has three bands.		(Esquinas Alcazar, 1981)
Pgm-1	*PGM-21* *Pgm-21*	*Phosphoglucomutase-1.* Isozyme variant with two alleles, each regulating two bands. The heterozygotes has three bands.		(Esquinas Alcazar, 1981)
Pgm-2	*Pgm*	*Phosphoglucomutase.* Isozyme variant with two alleles, each regulating one band, in PI 218070, PI 179923. Relationship with *Pgm-1* is unknown.	A	(Staub et al., 1998)
Px-1	*PRX-11*	*Peroxidase-1.* Isozyme variant with two codominant alleles, each regulating a cluster of four adjacent bands. The heterozygote has five bands.		(Esquinas Alcazar, 1981)
Px-2	*Px2A* *Prx2*	Peroxidase-2. Isozyme variant with two codominant alleles, each regulating a cluster of three adjacent bands. The heterozygote has four bands.		(Dane, 1983; Chen et al., 1990)
Skdh-1	-	*Shikimate dehydrogenase-1.* Isozyme variant with two codominant alleles, each regulating one band. The heterozygote has three bands.		Chen et al., 1990) (Gang and Lee, 1998)

[z]Linkage groups to which the genes belong are indicated as letters, according to Staub et al. (1998), and Roman numbers according to Périn et al. (2002).

Table 3. Quantitative traits loci, including description of the quantitative trait, number of QTL reported, parental lines of the cross used, and references.

Description of the quantitative trait, parental lines of the cross used	References
Aphis gossypii resistance Four additive and two couples of epistatic QTL affecting behaviour and biotic potential of *Aphis gossypii* in the cross Védrantais x PI 161375 (RILs).	(Boissot et al., 2010)
Bemisia tabaci resistance Two QTL affecting the biotic potential of the whiteflies in the cross Védrantais x PI 161375 (RILs).	(Boissot et al., 2010)
Cucumber mosaic virus resistance. Seven QTL are involved in resistance to three different CMV strains in the cross Védrantais x PI 161375 (RILs). A single QTL required for controlling CMV P9 and P104.82 strains in the cross Piel de Sapo x PI 161375 (LG XII).	(Dogimont et al., 2000) (Essafi et al., 2009)
Fusarium oxysporum f.sp. *melonis* race 1.2 resistance Nine QTL described in the cross Védrantais x Isabelle.	(Perchepied et al., 2005)
Pseudoperonospora cubensis resistance Nine QTL for resistance to downy mildew described in the cross Védrantais x PI 124112.	(Perchepied et al., 2005)
Podosphaera xanthii resistance Two QTL for resistance to powdery mildew described in the cross TGR-1551 x Bola de Ora (F$_2$), a major one, dominant (LG V) and a minor one, recessive (LG VIII).	(Yuste-Lisbona et al., 2011)
Ovary shape Six QTL for ovary length, eight QTL for ovary width and six QTL for the ratio ovary length/ovary width described in the cross Védrantais x PI 161375 (RILs). Five QTL for ovary shape in the cross Piel de Sapo x PI 161375 (NILs).	(Perin et al., 2002) (Eduardo et al., 2007)
Fruit shape Four QTL for fruit length, 5 QTL for fruit width and 6 QTL for the ratio fruit length/fruit width described in the cross Védrantais x PI 161375. Four QTL for fruit length, one for fruit width and two for the ratio fruit length : fruit width described in the cross Védrantais x PI 414723, which are common to both crosses. Eight QTL for fruit shape described in the cross Piel de Sapo x PI 161375 (F$_2$ and DHLs). Eleven QTL for fruit length, 10 QTL for fruit width and 15 QTL for the ratio fruit length/fruit width described in the cross Piel de Sapo x PI 161375 (NILs). Two QTL for fruit length, 2 QTL for fruit width and QTL for the ratio fruit length/fruit width described in the PI 414723 x Dulce (RI).	(Perin et al., 2002) (Perin et al., 2002) (Monforte et al., 2004) (Eduardo et al., 2007; (Fernandez-Silva et al., 2010) (Harel-Beja et al., 2010)
Fruit weight Six QTL described in the cross Piel de Sapo x PI 161375 (F$_2$ and DHLs). Eleven QTL described in the cross Piel de Sapo x PI 161375 (NILs).	(Monforte et al., 2004) (Eduardo et al., 2007)
Fruit firmness Two QTL for fruit firmness of the whole fruit described in the cross PI 414723 x Dulce (RI).	(Harel-Beja et al., 2010)
Rind traits Three QTL for stripes, three QTL for sutures described in the cross PI 414723 x Dulce (RI).	(Harel-Beja et al., 2010)
External color of the fruit Four QTL described in the cross Piel de Sapo x PI 161375 (F$_2$ and DHLs). Four QTL described in the cross Piel de Sapo x PI 161375 (NILs). Thirteen QTL for skin color and 12 QTL for ground spot color using the three color components in the cross Piel de Sapo x PI 161375 (NILs).	(Monforte et al., 2004) (Eduardo et al., 2007) (Obando et al., 2008)
Flesh color Three QTL for orange flesh color described in the cross Piel de Sapo x PI 161375 (F$_2$ and DHLs). Four QTL for fruit flesh color described in the cross Piel de Sapo x PI 161375 (NILs). Sixteen QTL for flesh color and 10 QTL for juice color using the three color components in the cross Piel de Sapo x PI 161375 (NILs). Three QTL for flesh color described in the cross PI 414723 x Dulce (RI).	(Monforte et al., 2004) (Eduardo et al., 2007) (Obando et al., 2008) (Harel-Beja et al., 2010)

Description of the quantitative trait, parental lines of the cross used	References
Sugar content of fruit flesh in mature fruit Five QTL for soluble solid content described in the cross Piel de Sapo x PI 161375 (F_2 and DHLs). QTL for sucrose, total soluble solids in the cross TAM Dulce x TGR-1551 (F_2). Fifteen QTL for soluble solid content in the cross Piel de Sapo x PI 161375 (NILs). Twenty-seven QTL for sugars, eight for fructose, six for glucose, four for sucrose, nine for sucrose equivalents in the cross Piel de Sapo x PI 161375 (NILs). Six QTL for sucrose, total soluble solids in the cross Deltex x TGR-1551 (F_2). Six QTL for sucrose, total soluble solids in the cross PI 414723 x Dulce (RI).	(Monforte et al., 2004) (Sinclair et al., 2006) (Eduardo et al., 2007) (Obando-Ulloa et al., 2009) (Park et al., 2009) (Harel-Beja et al., 2010)
Organic acid profile of fruit flesh in mature fruit Twenty-one QTL for organic acids in the cross Piel de Sapo x PI 161375 (NILs).	(Obando-Ulloa et al., 2009)
Ascorbic acid One QTL in the in the cross Deltex x TGR-1551 (F_2).	(Park et al., 2009)
Ethylene production in fruit (climacteric crisis). Four QTL described in the cross Védrantais x PI 161375 (RILs). One QTL for ethylene production and climacteric response in the cross Piel de Sapo x PI 161375 (NILs), non-climacteric parental lines.	(Perin et al., 2002) (Moreno et al., 2008)
Fruit flesh firmness Five QTL for flesh firmness in the cross Piel de Sapo x PI 161375 (NILs).	(Moreno et al., 2008)
Fruit flesh arroma profile *Ester 3-hydroxy-2,4,4-trimethyl-pentyl 2-methylpropanoate:* Two QTL in the cross Piel de Sapo x PI 161375 (NILs). *(Z,Z)-3,6 nonadiena,* responsible for the cucumber-like aroma: One QTL in the cross Piel de Sapo x PI 161375 (NILs). *Octanal:* One QTL in the cross Piel de Sapo x PI 161375 (NILs).	(Obando-Ulloa et al., 2010)
Root growth and architecture Seventeen QTL for root traits in the cross Piel de Sapo x PI 161375 (NILs).	(Fita et al., 2008)
Earliness. Nine QTL described in the cross Piel de Sapo x PI 161375 (F_2 and DHLs). Three QTL for early fruit maturity in the cross Chinese line Q 3-2-2 x Top Mark (F_2-F_3).	(Monforte et al., 2004) (Cuevas et al., 2009)
Yield-related traits Four QTL for primary branch number, five QTL for fruit number per plant, four QTL for fruit weight per plant, two QTL for average weight per fruit and one QTL for percentage of mature fruit per plot in the cross USDA 846-1 x Top Mark (RILs).	(Zalapa et al., 2007)
Postharvest life traits Three QTL involved in reduced postharvest losses and 11 with a detrimental effect on fruits after storage in the cross Piel de Sapo x PI 161375 (NILs).	(Fernandez-Trujillo et al., 2007)
Sensory traits Thirty-two QTL including global appreciation, sweetness, sourness in the cross Piel de Sapo x PI 161375 (NILs).	(Obando-Ulloa et al., 2009)

Acknowledgements: The author would like to thank Dr. James McCreight for his critical review of this gene list.

Literature cited

Alvarez JM, Gonzalez-Torres R, Mallor C, Gomez-Guillamon ML (2005) Potential sources of resistance to Fusarium wilt and powdery mildew in melons. Hortscience 40: 1657-1660

Anagnostou K, Jahn M, Perl-Treves R (2000) Inheritance and linkage analysis of resistance to zucchini yellow mosaic virus, watermelon mosaic virus, papaya ringspot virus and powdery mildew in melon. Euphytica 116: 265-270

Anagnostou K, Kyle M (1996) Genetic relationships among resistance to zucchini yellow mosaic, watermelon mosaic virus, papaya ringspot virus, and powdery mildew in melon (*Cucumis melo*). Hortscience 31: 913-914

Angelov D, Krasteva L (2000) Dominant inheritance of downy mildew resistance in melons. *In* Proceedings of Cucurbitaceae 2000, pp 273-275

Arumanagathan K, Earle ED (1991) Nuclear DNA content of some important plant species. Plant Molecular Biology Reporter 9: 208-218

Arzani A, Ahoonmanesh A (2000) Study of resistance to cucumber mosaic virus, watermelon mosaic virus and zucchini mosaic virus in melon cultivars. Iran Agricultural Research 19: 129-144

Bains MS, Kang US (1963) Inheritance of some flower and fruit characters in muskmelon. Indian Journal of Genetics and Plant Breeding 23: 101-106

Baudracco-Arnas S, Pitrat M (1996) A genetic map of melon (*Cucumis melo* L) with RFLP, RAPD, isozyme, disease resistance and morphological markers. Theoretical and Applied Genetics 93: 57-64

Besombes D, Giovinazzo N, Olivier C, Dogimont C, Pitrat M (1999) Description and inheritance of an *albino* mutant in melon. Cucurbit Genetics Cooperative Rep. 22: 14-15

Bohn GW (1961) Inheritance and origin of nectarless muskmelon. Journal of Heredity 52: 233-237

Bohn GW (1968) A red stem pigment in muskmelon. Vegetable Improvement Newsletter 10: 107

Bohn GW, Kishaba AN, Principe JA, Toba HH (1973) Tolerance to melon aphid in *Cucumis melo* L. Journal of the American Society for Horticultural Science 98: 37-40

Bohn GW, Principe JA (1964) A second male-sterility gene in the muskmelon. Journal of Heredity 55: 211-215

Bohn GW, Whitaker TW (1949) A gene for male sterility in the muskmelon (*Cucumis melo* L.). Proceedings American Society Horticultural Science 53: 309-314

Bohn GW, Whitaker TW (1964) Genetics of resistance to powdery mildew race 2 in muskmelon. Phytopathology 54: 587-591

Boissot N, Thomas S, Sauvion N, Marchal C, Pavis C, Dogimont C (2010) Mapping and validation of QTLs for resistance to aphids and whiteflies in melon. Theoretical and Applied Genetics 121:9-20

Boualem A, Fergany M, Fernandez R, Troadec C, Martin A, Morin H, Sari MA, Collin F, Flowers JM, Pitrat M, Purugganan MD, Dogimont C, Bendahmane A (2008) A conserved mutation in an ethylene biosynthesis enzyme leads to andromonoecy in melons. Science 321: 836-838

Brotman Y, Kovalski I, Dogimont C, Pitrat M, Portnoy V, Katzir N, Perl-Treves R (2005) Molecular markers linked to papaya ring spot virus resistance and Fusarium race 2 resistance in melon. Theoretical and Applied Genetics 110: 337-345

Brotman Y, Silberstein L, Kovalski I, Perin C, Dogimont C, Pitrat M, Klingler J, Thompson GA, Perl-Treves R (2002) Resistance gene homologues in melon are linked to genetic loci conferring disease and pest resistance. Theoretical and Applied Genetics 104: 1055-1063

Burger Y, Sa'ar U, Distelfeld A, Katzir N, Yeselson Y, Shen S, Schaffer AA (2003) Development of sweet melon (*Cucumis melo*) genotypes combining high sucrose and organic acid content. Journal of the American Society for Horticultural Science 128: 537-540

Burger Y, Saar U, Katzir N, Paris HS, Yeselson Y, Levin I, Schaffer AA (2002) A single recessive gene for sucrose accumulation in *Cucumis melo* fruit. Journal of the American Society for Horticultural Science 127: 938-943

Chadha ML, Nandpuri KS, Singh S (1972) Inheritance of some fruit characters in muskmelon. Indian Journal of Horticulture 29: 58-62

Chen FC, Hsiao CH, Chang YM, Li HW (1990) Isozyme variation in *Cucumis melo* L. I. Peroxidase and shikimate dehydrogenase variation in four melon varieties and its application for F_1 hybrid identification. Journal Agricultural Research China 39: 182-189

Chikh-Rouhou H, Gonzalez-Torres R, Alvarez JM, Oumouloud A (2010) Screening and morphological characterization of melons for resistance to *Fusarium oxysporum* f.sp *melonis* race 1.2. Hortscience 45: 1021-1025

Chikh-Rouhou H, Torres RG, Alvarez JM (2008) Characterization of the resistance to *Fusarium oxysporum* f.sp *melonis* race 1.2 in *Cucumis melo* 'BG-5384'. *In* M Pitrat ed, Cucurbitaceae 2008: Proceedings of the Ixth Eucarpia Meeting on Genetics and Breeding of Cucurbitaceae, pp 419-422

Clayberg CD (1992) Interaction and linkage tests of flesh color genes in *Cucumis melo* L. Cucurbit Genetics Cooperative Rep. 15: 53

Cohen Y, Cohen S, Eyal H, Thomas CE (1985) Inheritance of resistance to downy mildew in *Cucumis melo* PI 124111. Cucurbit Genetics Cooperative Rep. 8: 36-38

Cohen Y, Meron I, Mor N, Zuriel S (2003) A new pathotype of *Pseudoperonospora cubensis* causing downy mildew in cucurbits in Israel. Phytoparasitica 31: 458-466

Committee CGL (1982) Update of cucurbit gene list and nomenclature rules. Cucurbit Genetics Cooperative Rep. 5: 62-66

Coudriet DL, Kishaba AN, Bohn GW (1981) Inheritance of resistance to muskmelon necrotic spot virus in a melon aphid-resistant breeding line of muskmelon. Journal of the American Society for Horticultural Science 106: 789-791

Cox EL (1985) Three new seedling marker mutants in *Cucumis melo*. Hortscience 20: 657-657

Cox EL, Harding KE (1986) Linkage relationships of the light-green mutant in cantaloupe. Hortscience 21: 940-940

Cuevas HE, Staub JE, Simon PW (2010) Inheritance of beta-carotene-associated mesocarp color and fruit maturity of melon (*Cucumis melo* L.). Euphytica 173: 129-140

Cuevas HE, Staub JE, Simon PW, Zalapa JE (2009) A consensus linkage map identifies genomic regions controlling fruit maturity and beta-carotene-associated flesh color in melon (*Cucumis melo* L.). Theoretical and Applied Genetics 119: 741-756

Cuevas HE, Staub JE, Simon PW, Zalapa JE, McCreight JD (2008) Mapping of genetic loci that regulate quantity of beta-carotene in fruit of US Western Shipping melon (*Cucumis melo* L.). Theoretical and Applied Genetics 117: 1345-1359

Dane F (1983) Cucurbit. *In* SD Tanksley, TJ Orton eds, Isozymes in plant genetics and breeding, part B, Elsevier Science Publication, Amsterdam (NL), pp 369-390

Danin-Poleg Y, Tadmor Y, Tzuri G, Reis N, Hirschberg J, Katzir N (2002) Construction of a genetic map of melon with molecular markers and horticultural traits, and localization of genes associated with ZYMV resistance. Euphytica 125: 373-384

Danin-Poleg Y, Tzuri G, Reis N, Karchi Z, Katzir N (2000) Search for molecular markers associated with resistance to viruses in melon. *In* Proceedings of Cucurbitaceae 2000, pp 399-403

Danin-Poleg Y, Paris HS, Cohen S, Rabinowitch HD, Karchi Z (1997) Oligogenic inheritance of resistance to zucchini yellow mosaic virus in melons. Euphytica 93: 331-337

Daryono BS, Somowiyarjo S, Natsuaki KT (2003) New source of resistance to cucumber mosaic virus in melon. SABRAO Journal of Breeding and Genetics 35: 19-26

Daryono BS, Wakui K, Natsuaki KT (2010) Linkage analysis and mapping of SCAR markers linked to CMV-B2 resistance gene in melon. Sabrao Journal of Breeding and Genetics 42: 35-45

Davis, R.M. (1970) Vein tracts not sutures in cantaloupe. HortScience 5:86

Denna DW (1962) A study of the genetic, morphological and physiological basis for the bush and vine habit of several cucurbits. Cornell University, Ithaca (NY, US)

Dhillon NPS, Ranjana R, Singh K, Eduardo I, Monforte AJ, Pitrat M, Dhillon NK, Singh PP (2007) Diversity among landraces of Indian snapmelon (*Cucumis melo* var. *momordica*). Genetic Resources and Crop Evolution 54: 1267-1283

Diaz-Pendon JA, Fernandez-Munoz R, Gomez-Guillamon ML, Moriones E (2005) Inheritance of resistance to Watermelon mosaic virus in *Cucumis melo* that impairs virus accumulation, symptom expression, and aphid transmission. Phytopathology 95: 840-846

Díaz A, Fergani M, Formisano G, Ziarsolo P, Blanca J, Fei Z, Staub JE, Zalapa JE, Cuevas HE, Dace G, Oliver M, Boissot N, Dogimont C, Pitrat M, Hofstede R, van Koert P, Harel-Beja R, Tzuri G, Portnoy V, Cohen S, Schaffer A, Katzir N, Xu Y, Zhang H, Fukino N, Matsumoto S, Garcia-Mas J, Monforte AJ (2011) A consensus linkage map for molecular markers and Quantitative Trait Loci associated with economically important traits in melon (*Cucumis melo* L.). BMC Plant Biology 11: 111

Diaz JA, Mallor C, Soria C, Camero R, Garzo E, Fereres A, Alvarez JM, Gomez-Guillamon ML, Luis-Arteaga M, Moriones E (2003) Potential sources of resistance for melon to nonpersistently aphid-borne viruses. Plant Disease 87: 960-964

Dogimont C, Bendahmane A, Chovelon V, Boissot N (2010) Host plant resistance to aphids in cultivated crops: Genetic and molecular bases, and interactions with aphid populations. Comptes Rendus Biologies 333: 566-573

Dogimont C, Bendahmane A, Pitrat M, Burget-Bigeard E, Hagen L, Le Menn A, Pauquet J, Rousselle P, Caboche M, Chovelon V (2004) New polynucleotide implicated in plant resistance, useful for producing transgenic plants resistant to *Aphis gossypii* and association viral transmission, also encoded protein. World patent WO2004/072109-A1, France

Dogimont C, Bordat D, Pages C, Boissot N, Pitrat M (1999) One dominant gene conferring the resistance to the leafminer, *Liriomyza trifolii* (Burgess) Diptera : Agromyzidae in melon (*Cucumis melo* L.). Euphytica 105: 63-67

Dogimont C, Bussemakers A, Martin J, Slama S, Lecoq H, Pitrat M (1997) Two complementary recessive genes conferring resistance to cucurbit aphid borne yellows luteovirus in an Indian melon line (*Cucumis melo* L.). Euphytica 96: 391-395

Dogimont C, Leconte L, Perin C, Thabuis A, Lecoq H, Pitrat M (2000) Identification of QTLs contributing to resistance to different strains of cucumber mosaic cucumovirus in melon. Proceedings of Cucurbitaceae 2000: 391-398

Dogimont C, Slama S, Martin J, Pitrat M (1996) Sources of resistance to cucurbit aphid-borne yellows luteovirus in a melon germ plasm collection. Plant Disease 80: 1379-1382

Dyutin KE (1967) (A spontaneous melon mutant with dissected leaves) (in Russian). Genetica 9: 179-180

Dyutin KE (1979) (Inheritance of yellow-green coloration of the young leaves in melon) (in Russian). Tsitologia i genetika 13: 407-408

Eduardo I, Arus P, Monforte AJ, Obando J, Fernandez-Trujillo JP, Martinez JA, Alarcon AL, Alvarez JM, van der Knaap E (2007) Estimating the genetic architecture of fruit quality traits in melon using a genomic library of near isogenic lines. Journal of the American Society for Horticultural Science 132: 80-89

Enzie WD (1943) A source of muskmelon mosaic resistance found in the oriental pickling melon, *Cucumis melo* var. *conomon*. Proceedings American Society Horticultural Science 43: 195-198

Epinat C, Pitrat M (1989) Inheritance of resistance of three lines of muskmelon (*Cucumis melo*) to downy mildew (*Pseudoperonospora cubensis*). *In* CE Thomas (ed), 'Cucurbitaceae 89', Charleston (SC, US), pp 133-135

Epinat C, Pitrat M, Bertrand F (1993) Genetic analysis of resistance of 5 melon lines to powdery mildews. Euphytica 65: 135-144

Esquinas Alcazar JT (1975) 'Hojas hendidas', a nuevo mutante en *Cucumis melo* L. Inst. Nacionale Investigaciones Agrarias An; Ser.: Produc. Veg. 5: 93-103

Esquinas Alcazar JT (1981) Allozyme variation and relationships among Spanish land races of *Cucumis melo* L. Kulturpflanze 29: 337-352

Essafi A, Diaz-Pendon JA, Moriones E, Monforte AJ, Garcia-Mas J, Martin-Hernandez AM (2009) Dissection of the oligogenic resistance to Cucumber mosaic virus in the melon accession PI 161375. Theoretical and Applied Genetics 118: 275-284

Esteva J, Nuez F (1992) Tolerance to a whitefly-transmitted virus causing muskmelon yellows disease in Spain. Theoretical and Applied Genetics 84: 693-697

Esteva J, Nuez F, Gomez-Guillamon ML (1989) Resistance to yellowing disease in muskmelon. Cucurbit Genetic Cooperative Report 12: 44-45

Fergany M, Kaur B, Monforte AJ, Pitrat M, Rys C, Lecoq H, Dhillon NPS, Dhaliwal SS (2011) Variation in melon (Cucumis melo) landraces adapted to the humid tropics of southern India. Genetic Resources and Crop Evolution 58: 225-243

Fernandez-Silva I, Moreno E, Essafi A, Fergany M, Garcia-Mas J, Martin-Hernandez AM, Alvarez JM, Monforte AJ (2010) Shaping melons: agronomic and genetic characterization of QTLs that modify melon fruit morphology. Theoretical and Applied Genetics 121: 931-940

Fernandez-Trujillo JP, Obando J, Martinez JA, Alarcon AL, Eduardo I, Arus P, Monforte AJ (2007) Mapping fruit susceptibility to postharvest physiological disorders and decay using a collection of near-isogenic lines of melon. Journal of the American Society for Horticultural Science 132: 739-748

Fita A, Pico B, Monforte AJ, Nuez F (2008) Genetics of root system architecture using near-isogenic lines of melon. Journal of the American Society for Horticultural Science 133: 448-458

Foster RE (1963) Glabrous, a new seedling marker in muskmelon. Journal of Heredity 54: 113-114

Foster RE, Bond WT (1967) Abrachiate, an androecious mutant muskmelon. Journal of Heredity 58: 13-14

Frantz JD, Jahn MM (2004) Five independent loci each control monogenic resistance to gummy stem blight in melon (Cucumis melo L.). Theoretical and Applied Genetics 108: 1033-1038

Fukino N, Ohara T, Monforte AJ, Sugiyama M, Sakata Y, Kunihisa M, Matsumoto S (2008) Identification of QTLs for resistance to powdery mildew and SSR markers diagnostic for powdery mildew resistance genes in melon (Cucumis melo L.). Theoretical and Applied Genetics 118: 165-175

Fukino N, Ohara T, Sakata Y, Kunihisa M, Matsumoto S (2006) Quantitative trait locus analysis of powdery mildew resistance against two strains of Podosphaera xanthii in the melon line 'PMAR No. 5'. In GJ Holmes (ed), Proceedings of Cucurbitaceae 2006. Universal Press, Raleigh (NC, US), Asheville (NC, US), pp 95-99

Gabillard D, Pitrat M (1988) A fasciated mutant in Cucumis melo. Cucurbit Genetics Cooperative Rep. 11: 37-38

Galperin M, Zelcer A, Kenigsbuch D (2003) High competence for adventitious regeneration in the BU-21/3 melon genotype is controlled by a single dominant locus. Hortscience 38: 1167-1168

Ganesan J (1988) Genetic studies on certain characters of economic importance in muskmelon (Cucumis melo L.). Annamalai University (India)

Ganesan J, Sambandam CN (1979) Inheritance of certain qualitative characters in muskmelon (Cucumis melo L.). Annamalai University Agricultural Research Annals 9: 41-44

Ganesan J, Sambandam CN (1985) Inheritance of leaf shape in muskmelon (Cucumis melo L.) I. A qualitative approach. Annamalai University Agricultural Research Annals 12: 53-58

Gang T, Lee J (1998) Isozyme analysis and its application for purity test of F_1 hybrid seeds in melons. Journal of the Korean Society for Horticultural Science 39: 266-272

Gilbert RZ, Kyle MM, Munger HM, Gray SM (1994) Inheritance of resistance to watermelon mosaic-virus in Cucumis melo L. Hortscience 29: 107-110

Gómez-Guillamón ML, López-Sesé AI, Sarria E, Yuste-Lisbona FJ (2006) Linkage analysis among resistances to powdery mildew and virus transmission by Aphis gossypii Glover in melon line 'TGR-1551'. In GJ Holmes (ed), Proceedings of Cucurbitaceae 2006. Universal Press, Raleigh (NC, US), Asheville (NC, US), pp 100-107

Gómez-Guillamón ML, Moriones E, Luís-Arteaga M, Alvarez JM, Torés JA, López-Sesé AI, Cánovas I, Sánchez F, Camero R (1998) Morphological and disease resistances evaluation in Cucumis melo and its wild relatives. In JD McCreight (ed), Cucurbitaceae '98 Evaluation and enhancement of Cucurbits germplasm. ASHS Press, Alexandria (VA, USA), Pacific Grove (CA, US), pp 53-61

Gonzalez-Garza R, Gumpf DJ, Kishaba AN, Bohn GW (1979) Identification, seed transmission, and host range pathogenicity of a California isolate of melon necrotic spot virus. Phytopathology 69: 340-345

Gonzalez VM, Garcia-Mas J, Arus P, Puigdomenech P (2010) Generation of a BAC-based physical map of the melon genome. BMC Genomics 11

Gray SM, Moyer JW, Kennedy GG (1988) Resistance in Cucumis melo to watermelon mosaic virus-2 correlated with reduced virus movement within leaves. Phytopathology 78: 1043-1047

Hagiwara T, Kamimura K (1936) Cross-breeding experiments in Cucumis melo. Tokyo Horticultural School Publication

Harel-Beja R, Tzuri G, Portnoy V, Lotan-Pompan M, Lev S, Cohen S, Dai N, Yeselson L, Meir A, Libhaber SE, Avisar E, Melame T, van Koert P, Verbakel H, Hofstede R, Volpin H, Oliver M, Fougedoire A, Stalh C, Fauve J, Copes B, Fei Z, Giovannoni J, Ori N, Lewinsohn E, Sherman A, Burger J, Tadmor Y, Schaffer AA, Katzir N (2010) A genetic map of melon highly enriched with fruit quality QTLs and EST markers, including sugar and carotenoid metabolism genes. Theoretical and Applied Genetics 121: 511-533

Harwood RR, Markarian D (1968) A genetic survey of resistance to powdery mildew in muskmelon. Journal of Heredity 59: 213-217

Harwood RR, Markarian D (1968) The inheritance fo resistance to powdery mildew in the cantaloupe variety Seminole. Journal of Heredity 59: 126-130

Hassan AA, Al-Masri HH, Obaji UA, Wafi MS, Quronfilah NE, Al-Rays MA (1991) Screening of domestic and wild Cucumis melo germplasm for resistance to the yellow-stunting disorder in the United Arab Emirates. Cucurbit Genetic Cooperative Report 14: 56-58

Hassan AA, Merghany MM, Abdel-Ati KA, Abdel-Salam AM, Ahmed YM (1998) Inheritance of resistance to interveinal mottling and yellowing disease in cucurbits. Egyptian Journal of Horticulture 25: 209-224

Herman R, Perl-Treves R (2007) Characterization and inheritance of a new source of resistance to Fusarium oxysporum f. sp melonis race 1.2 in Cucumis melo. Plant Disease 91: 1180-1186

Herman R, Zvirin Z, Kovalski I, Freeman S, Denisov Y, Zuri G, Katzir N, Perl-Treves R (2008) Characterization of Fusarium race 1.2 resistance in melon and mapping of a major QTL for this trait near a fruit netting locus. In M Pitrat ed,

Cucurbitaceae 2008: Proceedings of the Ixth Eucarpia Meeting on Genetics and Breeding of Cucurbitaceae, pp 149-156

Hirai S, Amemiya Y (1989) Studies on the resistance of melon cultivars to cucumber mosaic virus. I Virus multiplication in leaves or mesophyll protoplasts from susceptible and resistant cultivars. Annals of the Phytopathological Society of Japan 55: 458-465

Hoffman JC, Nugent PE (1973) Inheritance of a virescent mutant of muskmelon. Journal of Heredity 64: 311-312

Hughes MB (1948) The inheritance of two characters of *Cucumis melo* and their interrelationship. Proceedings American Society Horticultural Science 52: 399-402

Iglesias A, Pico B, Nuez F (2000) A temporal genetic analysis of disease resistance genes: resistance to melon vine decline derived from *Cucumis melo* var. *agrestis*. Plant Breeding 119: 329-334

Iman MK, Abo-Bakr MA, Hanna HY (1972) Inheritance of some economic characters in crosses between sweet melon and snake cucumber. I. Inheritance of qualitative characters. Assiut Journal Agricultual Science 3: 363-380

Jagger IC, Whitaker TW, Porter DR (1938) Inheritance in *Cucumis melo* of resistance to powdery mildew (*Erysiphe cichoracearum*). Phytopathology 28: 761

Jolley VD, Brown JC, Nugent PE (1991) A genetically related response to iron-deficiency stress in muskmelon. Plant and Soil 130: 87-92

Joobeur T, King JJ, Nolin SJ, Thomas CE, Dean RA (2004) The fusarium wilt resistance locus *Fom-2* of melon contains a single resistance gene with complex features. Plant Journal 39: 283-297

Karchi Z, Cohen S, Govers A (1975) Inheritance of resistance to cucumber mosaic virus in melons. Phytopathology 65: 479-481

Kenigsbuch D, Cohen Y (1987) Inheritance of gynoecious sex type in muskmelon. Cucurbit Genetics Cooperative Rep. 10: 47-48

Kenigsbuch D, Cohen Y (1989) Independent inheritance of resistance to race-1 and race-2 of *Sphaerotheca fuliginea* in muskmelon. Plant Disease 73: 206-208

Kenigsbuch D, Cohen Y (1990) The inheritance of gynoecy in muskmelon. Genome 33: 317-320

Kenigsbuch D, Cohen Y (1992) Inheritance of resistance to downy mildew in *Cucumis melo* PI 124112 and commonality of resistance genes with PI 124111F. Plant Disease 76: 615-617

Kim H, Baek J, Choi YO, Lee JH, Sung SK, Kim S (2010) Identification of a cluster of oligonucleotide repeat sequences and its practical implication in melon (*Cucumis melo* L.) breeding. Euphytica 171: 241-249

Kishaba AN, Bohn GW, Toba HH (1971) Resistance to *Aphis gossypii* in muskmelon. Journal of Economic Entomology 64: 935-937

Kishaba AN, Bohn GW, Toba HH (1976) Genetic aspects of antibiosis to *Aphis gossypii* in *Cucumis melo* line from India. Journal of the American Society for Horticultural Science 101: 557-561

Knavel DE (1990) Inheritance of a short-internode mutant of Mainstream muskmelon. Hortscience 25: 1274-1275

Kubicki B (1962) Inheritance of some characters in muskmelons (*Cucumis melo*). Genetica Polonica 3: 265-274

Lebeda A, Kristkova E, Sedlakova B, Coffey MD, McCreight JD (2011) Gaps and perspectives of pathotype and race determination in *Golovinomyces cichoracearum* and *Podosphaera xanthii*. Mycoscience 52: 159-164

Lecouviour M, Pitrat M, Olivier C, Ricard M (1995) *Cochleare folium*, a mutant with spoon shaped leaf in melon. Cucurbit Genetics Cooperative Rep. 18: 37

Lecouviour M, Pitrat M, Risser G (1990) A fifth gene for male sterility in *Cucumis melo*. Cucurbit Genetics Cooperative Rep. 13: 34-35

Lee CW, Janick J (1978) Inheritance of seedling bitterness in *Cucumis melo* L. Hortscience 13: 193-194

Liu LZ, Chen YY, Su ZH, Zhang H, Zhu WM (2010) A sequence-amplified characterized region marker for a single, dominant gene in melon PI 134198 that confers resistance to a unique race of *Podosphaera xanthii* in China. Hortscience 45: 1407-1410

Lopez-Sese AI, Gomez-Guillamon ML (2000) Resistance to cucurbit yellowing stunting disorder virus (CYSDV) in *Cucumis melo* L. Hortscience 35: 110-113

Lozanov P (1983) Selekcija na mazkosterilni roditelski komponenti za ulesnjavana na proizvodstvoto na hibridni semena ot papesi. Dokl. na parva naucna konferencija po genetika i selekapa, Razgrad.

Lumsden D (1914) Mendelism in melons. New Hampshire Agricultural Experiment Station Bulletin 172: 58pp

Ma D, Sun L, Liu YH, Zhang Y, Liu H (1997) A genetic model of bitter taste in young fruits of melon. Cucurbit Genetics Cooperative Rep. 20: 27-29

Maestro-Tejada MC (1992) Résistance du melon aux virus. Interaction avec les pucerons vecteurs. Analyse génétique sur des lignées haplodiploïdes. Thèse de doctorat Université Aix Marseille III: 105p

Mallor C, Alvarez JM, Luis-Arteaga M (2003) A resistance to systemic symptom expression of melon necrotic spot virus in melon. Journal of the American Society for Horticultural Science 128: 541-547

Mallor Gimenez C, Álvarez JM, Luis Arteaga M (2003) Inheritance of resistance to systemic symptom expression of melon necrotic spot virus (MNSV) in *Cucumis melo* L. 'Doublon'. Euphytica 134: 319-324

Martin A, Troadec C, Boualem A, Rajab M, Fernandez R, Morin H, Pitrat M, Dogimont C, Bendahmane A (2009) A transposon-induced epigenetic change leads to sex determination in melon. Nature 461(7267): 1135-1138

McCreight JD (1983) Linkage of red stem and male sterile-1 in muskmelon. Cucurbit Genetics Cooperative Rep. 6: 48

McCreight JD (1983) A long internode mutant in muskmelon. Cucurbit Genetics Cooperative Rep. 6: 45

McCreight JD (1991) Potential sources of resistance to lettuce infectious yellows virus in melon. Cucurbit Genetic Cooperative Report 14: 51-52

McCreight JD (1992) Screening for lettuce infectious yellows virus resistance in melon. R.W. Doruchowski, E. Kozik and K. Niemirowicz-Szczytt, eds, Vth EUCARPIA Cucurbitaceae Symp., Poland: 160-162

McCreight JD (1998) Breeding melon for resistance to lettuce infectious yellows virus. J.D. McCreight, ed., Cucurbitaceae '98: Evaluation and enhancement of cucurbit germplasm: 241-247

McCreight JD (2000) Inheritance of resistance to lettuce infectious yellows virus in melon. Hortscience 35: 1118-1120

McCreight JD (2003) Genes for resistance to powdery mildew races 1 and 2US in melon PI 313970. Hortscience 38: 591-594

McCreight JD (2006) Melon-powdery mildew interactions reveal variation in melon cultigens and *Podosphaera xanthii* races 1 and 2. Journal of the American Society for Horticultural Science 131: 59-65

McCreight JD, Bohn GW (1979) Descriptions, genetics and independent assortment of red stem and pale in muskmelon (*Cucumis melo* L.). Journal of the American Society for Horticultural Science 104: 721-723

McCreight, J.D., Bohn, G.W., and Kishaba, A.N. 1992. 'Pedigree' of PI 414723 melon. Cucurbit Genet. Coop. Rpt. 15:51-52

McCreight JD, Coffey MD (2007) Resistance to a new race of the cucurbit powdery mildew present in Arizona and California. Hortscience 42: 1013-1013

McCreight JD, Coffey MD (2011) Inheritance of resistance in melon PI 313970 to cucurbit powdery mildew incited by *Podosphaera xanthii* race S. Hortscience 46: 838-840

McCreight JD, Elmstrom GW (1984) A third male-sterile gene in muskmelon. HortScience 19: 268-270

McCreight JD, Fashing-Burdette P (1996) Resistance of PI 124112 and 'Eldorado-300' melons (*Cucumis melo* L.) to papaya ringspot virus watermelon strain. *In* ML Gómez-Guillamón, C Soria, J Cuartero, JA Torès, R Fernandez-Munoz (eds), Cucurbits toward 2000. VIth EUCARPIA meeting on Cucurbit Genetics and Breeding, Málaga (ES), pp 298-301

McCreight JD, Kishaba AN, Bohn GW (1984) AR Hale's Best Jumbo, AR 5, and AR Topmark, melon aphid-resistant muskmelon breeding lines. HortScience 19: 309-310

McCreight JD, Liu HY, Turini TA (2008) Genetic resistance to Cucurbit leaf crumple virus in melon. Hortscience 43: 122-126

McCreight JD, Wintermantel WM (2008) Potential new sources of genetic resistance in melon to Cucurbit yellow stunting disorder virus. *In* Cucurbitaceae 2008: Proceedings of the Ixth Eucarpia Meeting on Genetics and Breeding of Cucurbitaceae, pp 173-179

Mockaitis JM, Kivilaan A (1965) A green corolla mutant in *Cucumis melo* L. Naturwissenschaften 52: 434

Mohamed ETI (1999) Collection and evaluation for disease resistance of melons from Sudan. PhD thesis, University of Gezira, Wad Medani, Sudan

Molina RV, Nuez F (1996) The inheritance of organogenic response in melon. Plant Cell Tissue and Organ Culture 46: 251-256

Monforte AJ, Oliver M, Gonzalo MJ, Alvarez JM, Dolcet-Sanjuan R, Arus P (2004) Identification of quantitative trait loci involved in fruit quality traits in melon (*Cucumis melo* L.). Theoretical and Applied Genetics 108: 750-758

Morales M, Luis-Arteaga M, Alvarez JM, Dolcet-Sanjuan R, Monfort A, Arus P, Garcia-Mas J (2002) Marker saturation of the region flanking the gene *NSV* conferring resistance to the melon necrotic spot Carmovirus (MNSV) in melon. Journal of the American Society for Horticultural Science 127: 540-544

Morales M, Orjeda G, Nieto C, van Leeuwen H, Monfort A, Charpentier M, Caboche M, Arus P, Puigdomenech P, Aranda MA, Dogimont C, Bendahmane A, Garcia-Mas J (2005) A physical map covering the *nsv* locus that confers resistance to Melon necrotic spot virus in melon (*Cucumis melo* L). Theoretical and Applied Genetics 111: 914-922

Moreno E, Obando JM, Dos-Santos N, Fernandez-Trujillo JP, Monforte AJ, Garcia-Mas J (2008) Candidate genes and QTLs for fruit ripening and softening in melon. Theoretical and Applied Genetics 116: 589-602

Moyer JW (1989) The effects of host resistance on viral pathogenesis in muskmelon. *In* CE Thomas (ed), 'Cucurbitaceae 89', Charleston (SC, US), pp 37-39

Moyer JW, Kennedy GG, Romanow LR (1985) Resistance to watermelon mosaic virus-II multiplication in *Cucumis melo*. Phytopathology 75: 201-205

Nieto C, Morales M, Orjeda G, Clepet C, Monfort A, Sturbois B, Puigdomenech P, Pitrat M, Caboche M, Dogimont C, Garcia-Mas J, Aranda MA, Bendahmane A (2006) An eIF4E allele confers resistance to an uncapped and non-polyadenylated RNA virus in melon. Plant Journal 48: 452-462

Nieto C, Piron F, Dalmais M, Marco CF, Moriones E, Gomez-Guillamon ML, Truniger V, Gomez P, Garcia-Mas J, Aranda MA, Bendahmane A (2007) EcoTILLING for the identification of allelic variants of melon eIF4E, a factor that controls virus susceptibility. BMC Plant Biology 7

Noguera FJ, Capel J, Alvarez JI, Lozano R (2005) Development and mapping of a codominant SCAR marker linked to the *andromonoecious* gene of melon. Theoretical and Applied Genetics 110: 714-720

Nuez F, Esteva J, Soria C, Gomez-Guillamon ML (1991) Search for sources of resistance to a whitefly transmitted yellowing disease in melon. Cucurbit Genetic Cooperative Report 14: 59-60

Nuez F, Pico B, Iglesias A, Esteva J, Juarez M (1999) Genetics of melon yellows virus resistance derived from *Cucumis melo* ssp *agrestis*. European Journal of Plant Pathology 105: 453-464

Nugent PE, Bhella HS (1988) A new chlorotic mutant of muskmelon. Hortscience 23: 379-381

Nugent PE, Cuthbert FP, Hoffman JC (1984) Two genes for cucumber beetle resistance in muskmelon. Journal of the American Society for Horticultural Science 109: 756-759

Nugent PE, Hoffman JC (1974) Inheritance of the halo cotyledon mutant in muskmelon. Journal of Heredity 65: 315-316

Obando-Ulloa JM, Eduardo I, Monforte AJ, Fernandez-Trujillo JP (2009) Identification of QTLs related to sugar and organic acid composition in melon using near-isogenic lines. Scientia Horticulturae 121: 425-433

Obando-Ulloa JM, Ruiz J, Monforte AJ, Fernandez-Trujillo JP (2010) Aroma profile of a collection of near-isogenic lines of melon (*Cucumis melo* L.). Food Chemistry 118: 815-822

Obando J, Fernandez-Trujillo JP, Martinez JA, Alarcon AL, Eduardo I, Arus P, Monforte AJ (2008) Identification of melon fruit quality quantitative trait loci using near-isogenic lines. Journal of the American Society for Horticultural Science 133: 139-151

Ohara T, Kojima A, Wako T, Ishiuchi D (2001) Inheritance of suppressed-branching in melon and its association with some other morphological characters. Journal of the Japanese Society for Horticultural Science 70: 341-345

Oumouloud A, Arnedo-Andres MS, Gonzalez-Torres R, Alvarez JM (2010) Inheritance of resistance to *Fusarium oxysporum* f. sp. *melonis* races 0 and 2 in melon accession Tortuga. Euphytica 176: 183-189

Oumouloud A, Torres RG, Andres MSA, Alvarez JM (2008) A new gene controlling resistance to *Fusarium oxysporum* f.sp *melonis* races 0 and 2 in melon. *In* M Pitrat ed, Cucurbitaceae 2008: Proceedings of the Ixth Eucarpia Meeting on Genetics and Breeding of Cucurbitaceae, pp 415-418

Paris HS, Nerson H, Karchi Z (1984) Genetics of internode length in melons. Journal of Heredity 75: 403-406

Park SO, Hwang HY, Crosby KM (2009) A genetic linkage map including loci for male sterility, sugars, and ascorbic acid in melon. Journal of the American Society for Horticultural Science 134: 67-76

Park SO, Hwang HY, Ham IK, Crosby KM (2009) Mapping of QTL controlling Ananas melon fruit net formation. Hortscience 44: 1145-1145

Parthasarathy VA, Sambandam CN (1981) Inheritance in Indian melons. Indian Journal of Genetics and Plant Breeding 41: 114-117

Pauquet J, Burget E, Hagen L, Chovelon V, A. LM, Valot N, Desloire S, Caboche M, Rousselle P, Pitrat M, Bendahmane A, Dogimont C (2004) Map-based cloning of the *Vat* gene from melon conferring resistance to both aphid colonization and aphid transmission of several viruses. *In* A Lebeda, H Paris (eds), Cucurbitaceae 2004, the 8th EUCARPIA meeting on Cucurbit genetics and breeding. Palaky University, Olomouc, Czech Republic, pp 325-329

Perchepied L, Bardin M, Dogimont C, Pitrat A (2005) Relationship between loci conferring downy mildew and powdery mildew resistance in melon assessed by quantitative trait loci mapping. Phytopathology 95: 556-565

Perchepied L, Dogimont C, Pitrat M (2005) Strain-specific and recessive QTLs involved in the control of partial resistance to *Fusarium oxysporum* f. sp *melonis* race 1.2 in a recombinant inbred line population of melon. Theoretical and Applied Genetics 111: 65-74

Perchepied L, Pitrat M (2004) Polygenic inheritance of partial resistance to *Fusarium oxysporum* f. sp *melonis* race 1.2 in melon. Phytopathology 94: 1331-1336

Périn C, Dogimont C, Giovinazzo N, Besombes D, Guitton L, Hagen L, Pitrat M (1999) Genetic control and linkages of some fruit characters in melon. Cucurbit Genetics Cooperative Rep. 22: 16-18

Perin C, Gomez-Jimenez M, Hagen L, Dogimont C, Pech JC, Latche A, Pitrat M, Lelievre JM (2002) Molecular and genetic characterization of a non-climacteric phenotype in melon reveals two loci conferring altered ethylene response in fruit. Plant Physiology 129: 300-309

Perin C, Hagen LS, De Conto V, Katzir N, Danin-Poleg Y, Portnoy V, Baudracco-Arnas S, Chadoeuf J, Dogimont C, Pitrat M (2002) A reference map of *Cucumis melo* based on two recombinant inbred line populations. Theoretical and Applied Genetics 104: 1017-1034

Perin C, Hagen LS, Giovinazzo N, Besombes D, Dogimont C, Pitrat M (2002) Genetic control of fruit shape acts prior to anthesis in melon (*Cucumis melo* L.). Molecular Genetics and Genomics 266: 933-941

Pitrat M (1978) Tolerance of melon to *Watermelon mosaic virus II*. Cucurbit Genetics Cooperative Rep. 1: 20

Pitrat M (1986) Gene list for muskmelon (*Cucumis melo* L.). Cucurbit Genetics Cooperative Rep. 9: 111-120

Pitrat M (1990) Gene list for *Cucumis melo* L. Cucurbit Genetics Cooperative Rep. 13: 58-68

Pitrat M (1991) Linkage groups in *Cucumis melo* L. Journal of Heredity 82: 406-411

Pitrat M (1994) Gene list for *Cucumis melo* L. Cucurbit Genetics Cooperative Rep. 17: 135-147

Pitrat M (1998) 1998 Gene list for melon. Cucurbit Genetics Cooperative Rep. 21: 69-81

Pitrat M (2002) 2002 Gene list for melon. Cucurbit Genetics Cooperative Rep. 25: 76-93

Pitrat M (2006) 2006 Gene list for melon. Cucurbit Genetics Cooperative Rep. 28-29: 142-163

Pitrat M, Besombes D (2008) Inheritance of *Podosphaera xanthii* resistance in melon line '90625'. *In* M Pitrat (ed), Cucurbitaceae 2008, IXth EUCARPIA meeting on Genetics and Breeding of Cucurbitaceae. INRA, Avignon (FRA), pp 135-142

Pitrat M, Ferrière C, Ricard M (1986) *Flava*, a chlorophyll deficient mutant in muskmelon. Cucurbit Genetics Cooperative Rep. 9: 67

Pitrat M, Lecoq H (1980) Inheritance of resistance to Cucumber MosaicVirus transmission by *Aphis gossypii* in *Cucumis melo*. Phytopathology 70: 958-961

Pitrat M, Lecoq H (1982) Genetic relations between non-acceptance and antibiosis resistance to *Aphis gossypii* in melon - Search for linkage with other genes. Agronomie 2: 503-507

Pitrat M, Lecoq H (1983) Two alleles for watermelon mosaic virus 1 resistance in melon. Cucurbit Genetics Cooperative: 52-53

Pitrat M, Lecoq H (1984) Inheritance of Zucchini Yellow Mosaic Virus resistance in *Cucumis melo* L. Euphytica 33: 57-61

Pitrat M, Olivier C, Ricard M (1995) A virescent mutant in melon. Cucurbit Genetics Cooperative Rep. 18: 37

Pitrat M, Risser G, Bertrand F, Blancard D, Lecoq H (1996) Evaluation of a melon collection for disease resistances. Cucurbits Towards 2000: 49-58

Pitrat M, Risser G, Ferrière C, Olivier C, Ricard M (1991) Two virescent mutants in melon (*Cucumis melo*). Cucurbit Genetics Cooperative Rep. 14: 45

Poole CF, Grimball PC (1939) Inheritance of new sex forms in *Cucumis melo* L. Journal of Heredity 30: 21-25

Prasad K, Norton JD (1967) Inheritance of resistance to *Mycosphaerella citrullina* in muskmelon. Proceedings American Society Horticultural Science 91: 396-400

Provvidenti R (1989) Sources of resistance to viruses in cucumber, melon, squash and watermelon. Proceedings of Cucurbitaceae 89 : Evaluation and Enhancement of Cucurbit Germplasm, Nov 29-Dec 2, 1989. Charleston, SC (USA). Ed. Thomas C.E.: 29-36

Provvidenti R (1993) Resistance to viral disease of cucurbits. M.M. Kyle, ed. Resistance to viral disease of vegetables. Timber Pres Portland, Oregon: 8-43

Provvidenti R (1998) A source of a high level of tolerance to squash mosaic virus in a melon from China. Cucurbit Genetic Cooperative Report 21: 29-30

Quiot-Douine L, Lecoq H, Quiot JB, Pitrat M, Labonne G (1988) Evidence for a biological and serological variability in a potyvirus infecting cucurbit: the *Papaya ringspot virus* (PRSV). *In* G Risser, M Pitrat (eds), EUCARPIA meeting 'Cucurbitaceae 88', Avignon (FR), pp 35-42

Quiot JB, Kaan F, Beramis M (1971) Identification d'une souche de la mosaïque de la pastèque (Watermelon Mosaic Virus 1) aux Antilles francaises. Annales de Phytopathologie 3: 125-130

Ray DT, McCreight JD (1996) Yellow-tip: A cytoplasmically inherited trait in melon (*Cucumis melo* L). Journal of Heredity 87: 245-247

Risser G (1973) Étude de l'hérédité de la résistance du melon (*Cucumis melo*) aux races 1 et 2 de *Fusarium oxysporum* f.sp. *melonis*. Annales de l'Amélioration des Plantes 23: 259-263

Risser G, Banihashemi Z, Davis DW (1976) Proposed nomenclature of *Fusarium oxysporum* f. sp. *melonis* races and resistance genes in *Cucumis melo*. Phytopathology 66: 1105-1106

Risser G, Pitrat M, Lecoq H, Rode JC (1981) Varietal susceptibility of melon to Muskmelon yellow stunt virus (Mysv) and to its transmission by *Aphis-Gossypii* - Inheritance of the wilting reaction. Agronomie 1: 835-838

Risser G, Pitrat M, Rode JC (1977) Resistance of melon (*Cucumis melo* L) to Cucumber Mosaic Virus. Annales de l'Amélioration des Plantes 27: 509-522

Rosa JT (1928) The inheritance of flower types in *Cucumis* and *Citrullus*. Hilgardia 3: 233-250.

Sambandam CN, Chelliah S (1972) *Cucumis callosus* (Rottl.) Logn., a valuable material for resistance breeding in muskmelons. *In* 3rd International Symposium Sub-tropical Horticulture, pp 63-68

Sensoy S, Demir S, Buyukalaca S, Abak K (2007) Response of Turkish melon genotypes to *Fusarium oxysporum* f. sp *melonis* race 1 determined by inoculation tests and RAPD markers. European Journal of Horticultural Science 72: 220-227

Silberstein L, Kovalski I, Brotman Y, Perin C, Dogimont C, Pitrat M, Klingler J, Thompson G, Portnoy V, Katzir N, Perl-Treves R (2003) Linkage map of *Cucumis melo* including phenotypic traits and sequence-characterized genes. Genome 46: 761-773

Sinclair JW, Park SO, Lester G, Yoo KS, Crosby K (2006) Identification and confirmation of RAPD markers and andromonoecious associated with quantitative trait loci for sugars in melon. Journal of the American Society for Horticultural Science 131: 360-371

Soria C, Sese AIL, Gomez-Guillamon ML (1996) Resistance mechanisms of *Cucumis melo* var *agrestis* against *Trialeurodes vaporariorum* and their use to control a closterovirus that causes a yellowing disease of melon. Plant Pathology 45: 761-766

Staub JE, Meglic V, McCreight JD (1998) Inheritance and linkage relationships of melon (*Cucumis melo* L.) isozymes. Journal of the American Society for Horticultural Science 123: 264-272

Sugiyama M, Ohara T, Sakata Y (2006) A new source of resistance to Cucumber green mottle mosaic virus in melon. Journal of the Japanese Society for Horticultural Science 75: 469-475

Sugiyama M, Ohara T, Sakata Y (2007) Inheritance of resistance to Cucumber green mottle mosaic virus in *Cucumis melo* L. 'Chang Bougi'. Journal of the Japanese Society for Horticultural Science 76: 316-318

Tadmor Y, Burger J, Yaakov I, Feder A, Libhaber SE, Portnoy V, Meir A, Tzuri G, Sa'ar U, Rogachev I, Aharoni A, Abeliovich H, Schaffer AA, Lewinsohn E, Katzir N (2010) Genetics of flavonoid, carotenoid, and chlorophyll pigments in melon fruit rinds. Journal of Agricultural and Food Chemistry 58: 10722-10728

Takada K, Kanazawa K, Takatuka K (1975) Studies on the breeding of melon for resistance to powdery mildew. II. Inheritance of resistance to powdery mildew and correlation of resistance to other characters. Bulletin of the Vegetable and Ornamental Crops Research Station A2: 11-31

Teixeira APM, da Silva Barreto FA, Aranha Camargo LEA (2008) An AFLP marker linked to the *Pm-1* gene that confers resistance to *Podosphaera xanthii* race 1 in *Cucumis melo*. Genetics and Molecular Biology 31: 547-550

Tezuka T, Waki K, Yashiro K, Kuzuya M, Ishikawa T, Takatsu Y, Miyagi M (2009) Construction of a linkage map and identification of DNA markers linked to *Fom-1*, a gene conferring resistance to *Fusarium oxysporum* f.sp *melonis* race 2 in melon. Euphytica 168: 177-188

Tezuka TTT, Waki K, Kuzuya M, Ishikawa T, Takatsu Y, Miyagi M (2011) Development of new DNA markers linked to the Fusarium wilt resistance locus *Fom-1* in melon. Plant Breeding 130: 261-267

Thomas CE, Cohen Y, McCreight JD, Jourdain EL, Cohen S (1988) Inheritance of resistance to downy mildew in *Cucumis melo*. Plant Disease 72: 33-35

Thomas CE, McCreight JD, Jourdain EL (1990) Inheritance of resistance to *Alternaria cucumerina* in *Cucumis melo* line MR-1. Plant Disease 74: 868-870

Vashistha RN, Choudhury B (1974) Inheritance of resistance to red pumpkin beetle in muskmelon. Sabrao Journal 6: 95-97

Velich I, Fulop I (1970) A new muskmelon type of cut leaf character. Zoldsegtermesztes 4: 107-112

Wall JR (1967) Correlated inheritance of sex expression and fruit shape in *Cucumis*. Euphytica 16: 199-208

Wang SW, Yang JH, Zhang MF (2011) Developments of functional markers for *Fom-2*-mediated fusarium wilt resistance based on single nucleotide polymorphism in melon (*Cucumis melo* L.). Molecular Breeding 27: 385-393

Webb RE (1979) Inheritance of resistance to watermelon mosaic virus-1 in *Cucumis melo* L. Hortscience 14: 265-266

Webb RE, Bohn GW (1962) Resistance to cucurbit viruses in *Cucumis melo* L. Phytopathology 52: 1221 (Abstract)

Whitaker TW (1952) Genetic and chlorophyll studies of a yellow-green mutant in muskmelon. Plant Physiology 27: 263-268

Whitaker TW, Davis GN (1962) Cucurbit, botany, cultivation and utilization. Interscience Publisher, New York (US), pp 250

Yuste-Lisbona FJ, Capel C, Capel J, Lozano R, Gomez-Guillamon ML, Lopez-Sese AI (2008) Conversion of an AFLP fragment into one dCAPS marker linked to powdery mildew resistance in melon. *In* Cucurbitaceae 2008: Proceedings of the Ixth Eucarpia Meeting on Genetics and Breeding of Cucurbitaceae, pp 143-148

Yuste-Lisbona FJ, Capel C, Gomez-Guillamon ML, Capel J, Lopez-Sese AI, Lozano R (2011) Codominant PCR-based markers and candidate genes for powdery mildew resistance in melon (*Cucumis melo* L.). Theoretical and Applied Genetics 122: 747-758

Yuste-Lisbona FJ, Capel C, Sarria E, Torreblanca R, Gomez-Guillamon ML, Capel J, Lozano R, Lopez-Sese AI (2011) Genetic linkage map of melon (*Cucumis melo* L.) and localization of a major QTL for powdery mildew resistance. Molecular Breeding 27: 181-192

Zalapa JE, Staub JE, McCreight JD, Chung SM, Cuevas H (2007) Detection of QTL for yield-related traits using recombinant inbred lines derived from exotic and elite US Western Shipping melon germplasm. Theoretical and Applied Genetics 114: 1185-1201

Zheng XY, Wolff DW (2000) Randomly amplified polymorphic DNA markers linked to fusarium wilt resistance in diverse melons. Hortscience 35: 716-721

Zheng XY, Wolff DW, Baudracco-Arnas S, Pitrat M (1999) Development and utility of cleaved amplified polymorphic sequences (CAPS) and restriction fragment length polymorphisms (RFLPs) linked to the *Fom-2* fusarium wilt resistance gene in melon (*Cucumis melo* L.). Theoretical and Applied Genetics 99: 453-463

Zheng XY, Wolff DW, Crosby KM (2002) Genetics of ethylene biosynthesis and restriction fragment length polymorphisms (RFLPs) of ACC oxidase and synthase genes in melon (*Cucumis melo* L.). Theoretical and Applied Genetics 105: 397-403

Zink FW (1977) Linkage of virescent foliage and plant-growth habit in muskmelon. Journal of the American Society for Horticultural Science 102: 613-615

Zink FW (1986) Inheritance of a greenish-yellow corolla mutant in muskmelon. Journal of Heredity 77: 363-363

Zink FW (1990) Inheritance of a delayed lethal mutant in muskmelon. Journal of Heredity 81: 210-211

Zink FW, Gubler WD (1985) Inheritance of resistance in muskmelon to fusarium wilt. Journal of the American Society for Horticultural Science 110: 600-604

Zuniga TL, Jantz JP, Zitter TA, Jahn MK (1999) Monogenic dominant resistance to gummy stem blight in two melon (*Cucumis melo*) accessions. Plant Disease 83: 1105-1107

2010-2011 CGC Membership Directory

Attard, Everaldo. University of Malta, Institute of Agriculture, Msida MSD2080 Malta. Email: everaldo. attard@um.edu.mt Interests: Research on the economic importance of Ecballium elaterium (squirting cucumber)

Boyhan, George E. UGA, Southeast District Coop. Extention, P.O. Box 8112, GSU, Statesboro GA 30460-8112 USA. Email: gboyhan@uga.edu Interests: pumpkin and watermelon breeding.

Bell, Duane. Rupp Seeds, Inc., 17919 County Rd. B, Wauseon OH 43567-9458 USA. Email: duaneb @ruppseeds.com

Call, Adam D. North Carolina State University, Dept Horticulture, Campus Box 7609, Raleigh NC 27695-7609 USA. Email: adcall@ncsu.edu

Cohen, Roni. ARO, Newe Ya'ar Research Center, P.O. Box 1021, Ramat Yishay 30095 Israel. Email: ronico@volcani.agri.gov.il Interests: Plant pathology; root and foliar diseases of cucurbits

Chung, Paul. Seminis Vegetable Seeds, Inc., 37437 State Highway 16, Woodland CA 95695 USA. Email: paul.chung@seminis.com Interests: Melon Breeding.

Crosby, Kevin. Texas A&M University, Vegetable & Fruit Improvement Ctr, 1500 Research Way, Ste 120, College Station TX 77845 USA. Email: k-crosby @tamu.edu Interests: Myrothecium stem canker on melon.

Davidi, Haim. Hazait 3, Moshav Beit Elazari 76803 Israel. Email: haimdavi@zahav.net.il

Davis, Angela R. USDA, ARS WWARL, P.O. Box 159, 911 E. Hwy 3, Lane OK 74555 USA. Email: angela.davis@ars.usda.gov, angela.davis@lane-ag.org Interests: Germplasm improvement.

de Groot, Erik. Nunhems Italy SRL, Via Ghiarone, 2; 40019 Sant' Agata, Bolognese BO Italy. Email: Erik.degroot@nunhems.com Interests: Watermelon breeding

de Hoop, Simon Jan. East-West Seed Co., 50/1 Moo 2, Sainoi-Bang Bua Thong Road, Sainoi Nonthaburi 11150 Thailand. Email: simon.dehoop@eastwestseed.com Interests: Cucurbit breeding

Dombrowski, Cory. 137 E. Lake Dr., Lehigh Acres FL 33936 USA. Email: cdombrowski@sakata.com Interests: cucurbit variety evaluation and melon breeding

Everts, Kathryne. 27664 Nanticoke Rd, Salisbury MD 21801 USA. Email: keverts@umd.edu Interests: Fusarium wilt on watermelon; Gummy stem blight on watermelon and other cucurbits; Cover crops for managing cucurbit diseases, fruit rot on pumpkin

Ficcadenti, Nadia. CRA-ORA, Unita di Ricera per l'Orticoltora, Via Salaria, 163030 Monsampolo del Tronto (A.P.) Italy. Email: nadiaf@insinet.it

Froese, David C. 32971 CR 34, La Junta CO 81050 USA. Email: david@coloradoseeds.com

Furuki, Toshi. Manager of Breeding Dept. 1, Kakegawa Research Center, Sakata Seed Corporation1743-2 Yoshioka, Kakegawa, 436-0115 Japan. Email: t.furuki@sakata-seed.co.jp

Garza-Ortega, Sergio. Plan de Iguala 66, Col. Mision del Sol, 83100 Hermosillo Sonora Mexico. Email: sgarzao@prodigy.net.mx Interests: Breeding of Cucurbita spp.; testing new muskmelon lines

Gardingo, Jose Raulindo. 142 Albert Einstein, Uvaranas, Ponta Grossa PR 84032-015 Brazil. Email: jrgardin@uepg.br

Gatto, Gianni. Esasem Spa, Via G. Marconi 56, 37052 Casaleone (VR) Italy. Email: ggatto@esasem.com; gpadox@gmail.com

Gil-Albert, Carlos. Rijk Zwaan Iberica, Att. Carlos Gil-Albert, Paraje El Mamí, Carretera De Viator s/n 04120 La Cañada, Almeria Almeria Spain. Email: c.gil@rijkzwaan.es

Goldman, Amy P. 164 Mountain View Road, Rhinebeck NY 12572 USA. Email: agoldthum@aol.com Interests: Heirloom melons and watermelons; ornamental gourds; garden writing

Grant, Doug. Hybrid Seed Co., 326 c Patumahoe Road, RD 3 Pukekohe, 2678 Auckland New Zealand. Email: doug.grant@xtra.co.nz Interests: Breeding & genetics of maxima and moschata

Grumet, Rebecca. Dept. of Horticulture, Graduate Program in Genetics, Michigan State University, East Lansing MI 48824-1325 USA. Email: grumet@msu.edu Interests: Disease resistance, gene flow, tissue culture and genetic engineering

Guner, Nihat. Sakata Seed America, P.O. Box 1118, Lehigh Acres FL 33970-1118 USA. Email: nguner @sakata.com Interests: Watermelon breeding

Haizhen, Li. Beijing Vegetable Research Center, P.O. Box 2443, Beijing 100097 P.R. China. Email: lihaizhen @nercv.com Interests: Cucurbita sp.

Havey, Michael J. USDA/ARS, Dept. of Horticulture, University of Wisconsin, 1575 Linden Dr., Madison WI 53706 USA. Email: mjhavey@wisc.edu

He, Xiaoming. Vegetable Research Institute; Guangdong Academy of Agric. Sciences; Guangzhou, Guangdong 510640; P.R. of CHINA; xiaominghe@tom.com; Breeding cucumber and wax gourd

Herrington, Mark. Maroochy Research Station, Dept Primary Industries & Fisheries, P.O. Box 5083, SCMC Nambour, Queensland 4560 Australia. Email: mark.herrington@dpi.qld.gov.au Interests: Cucurbita breeding

Hertogh, Kees. Nickerson-Zwaan BV, P.O. Box 28, 4920 AA, Made Netherlands. Email: kees.hertogh@nickerson-zwaan.com

Hofstede, Rene. Wageningen - Keygene N.V., Agro Business Park 90, Rabobank Vallei, 6701 Rijn, Wageningen Netherlands. Email: rho@keygene.com Interests: Molecular genetic research in all cucurbitaceae

Holman, Bohuslav. 1420 Bzinska Str., 69681 Bzenec Czech Republic. Email: bholman@iol.cz Interests: Cucumber breeding and seed production.

Hoogland, Jan. Bejo Zaden BV, Postbus 50, Trambaan 21749 ZH Warmenhuizen Netherlands. Email: j.hoogland@bejo.nl

Hu, Cheng-Jung. Crop Improvement Section, HDAIS, COA, R.O.C., 50, 2nd Sec, Chi-An Road, Chi-An Township, Hualien 97365 Taiwan. Email: greenficus @pchome.com.tw

Ignart, Frederic. Centre de Recherche CLAUSE TEZIER, Domaine de Maninet, Route de Beaumont, 26000 Va-

lence France. Email: frederic.ignart@clause-vegseeds.com Interests: melon breeding

Ito, Kimio. Vegetable Breeding Laboratory, Hokkaido National Agricultural Expt. Station, Hitsujigaoka Sapporo Japan. Email: kito@cryo.affrc.go.jp

Jahn, Laboratory of Molly. Dept. of Plant Breeding & Genetics, 312 Bradfield Hall, Ithaca NY 14853-1902 USA. Email: mjahn@cals.wisc.edu Interests: Melon and squash breeding and genetics

Johnson, Bill. Seminis Vegetable Seeds, 37237 State Hwy 16, Woodland CA 95695 USA. Email: bill.johnson @seminis.com Interests: Squash breeding

Johnston, Jr., Robert. Johnny's Selected Seeds, 184 Foss Hill Rd, Albion ME 04910-9731 USA. Email: rjohnston@johnnyseeds.com Interests: Squash and pumpkins

Jones-Evans, Elen. Peotec Seeds SRL, Via Provinciale 42-44, 43018 Sissa (PR) Italy. Email: ejevans @peotecseeds.com

Juarez, Benito. 2387 McNary Way, Woodland CA 95776 USA. Email: benito.juarez@alumni.ucdavis.edu Interests: Watermelon & melon genetics, breeding, physiology & postharvest

Kabelka, Eileen. Harris Moran, 9241 Mace, Woodland CA 95618 USA. Email: e.kabelka@hmclause.com

Karchi, Zvi. 74 Hashkedim St., Qiryat-Tiv'on 36501 Israel. Interests: Cucurbit breeding, cucurbit physiology

Katzir, Nurit. Newe Ya'ar Research Center, ARO, P.O. Box 1021, Ramat Yishay 30095 Israel. Email: katzirn@volcani.agri.gov.il

Kelfkens, Marcel. Westeinde 62, 1601 BK Enkhuizen Netherlands. Email: marcel.kelfkens@syngenta.com

King, Stephen R. Dept. of Horticultural Sciences, Texas A&M University, College Station TX 77843-2133 USA. Email: srking@tamu.edu Interests: Watermelon breeding

Kirkbride, Jr., Joseph H. U.S. National Arboretum, 3501 New York Ave. NE, Washington DC 20002-1958 USA. Email: joseph.kirkbride@ars.usda.gov Interests: Taxonomy of Cucumis

Knerr, Larry D. Shamrock Seed Company, 3 Harris Place, Salinas CA 93901-4586 USA. Email: lknerr @shamrockseed.com Interests: Varietal development of honeydew and cantaloupe

Kobori, Romulo Fujito. Av. Dr. Plínio Salgado, no 4320, Bairro Uberaba, CEP 12906-840Braganca Paulista Sao Paulo Brazil. Email: romulo.kobori@sakata.com.br

Kole, Chittaranjan. Clemson University, 100 Jordan Hall, Clemson SC 29634 USA. Email: ckole@clemson.edu Interests: Molecular mapping and breeding for phytomedicines in bitter melon

Kumar, Rakesh. 2336 Champion Court, Raleigh NC 27606 USA. Email: rklnu@ncsu.edu Interests: integration of conventional breeding with molecular techniques

Lanini, Brenda. Harris Moran Seed Co., 9241 Mace Blvd., Davis CA 95616 USA. Email: b.lanini@HMClause.com

Lebeda, Aleš. Faculty of Science, Dept. Botany, Palacky University, Slechtitelu 11783 71 Olomouc-Holice Moravia Czech Republic. Email: ales.lebeda@upol.cz; http://botany.upol.cz Interests: Cucurbitaceae family, genetic resources, diseases, fungal variability, resistance breeding, tissue culture

Lehmann, Louis Carl. Louie's Pumpkin Patch, Poppelvägen 6 B, SE-541 48 Skövde Sweden. Email: louis.lehmann@pumpkinpatch.se Interests: Cucurbita - testing of squash and pumpkin for use in Southern Sweden

Lelley, Tamas Univ. of Nat. Resources & Applied Life Sci, Dept. for AgroBiotech. IFA, Institute for Plant Prod. Biotechnology, Konrad Lorenz Str., 20A-3430 Tulln Austria. Email: tamas.lelley@boku.ac.at Interests: Cucurbita spp.

Levi, Amnon. U.S. Vegetable Laboratory, 2700 Savannah Highway, Charleston SC 29414 USA. Email: amnon.levi@ars.usda.gov

Lopez-Anido, Fernando. Universidad Nacional Rosario, CC 14, Zavalla S 2125 ZAA Argentina. Email: felopez@fcagr.unr.edu.ar Interests: Breeding of Cucurbita pepo L. (caserta type)

Liu, Wenge. Zhengzhou Fruit Research Inst., Chinese Acad. of Agric. Sci., Gangwan Rd 28, Guancheng District, Zhengzhou, Henan 450009 P.R. of China. Email: lwgwm@yahoo.com.cn Interests: Watermelon breeding, male sterility, tetraploids, triploids

Loy, J. Brent. Plant Biology Dept., G42 Spaulding Hall, Univ. of New Hampshire, 38 Academic Way, Durham NH 3824 USA. Email: jbloy@unh.edu Interests: Squash, melon, pumpkin. Genetics, breeding, plasticulture, mulch rowcovers

Luan, Feishi. Hort. College of Northeast Agri. Univ.; No. 59 Mucai St.; Xiangfang Dist., Heilongjiang 150030; P.R. of CHINA; luanfeishi@sina.com

Ma, Qing. College of Plant Protection, Northwest Agri. & Forestry University, Yangling, Shaanxi 712100 P.R. China. Email: maquing@nwsuaf.edu.cn Interests: Cucumber disease resistance, resistance mechanisms

Majde, Mansour. Gautier Semences, Route de'Avignon, 13630 Eyragues France. Email: mansour.majde @gautiersemences.com

Matsumoto, Yuichi. 3-18-10-203, Okazaki, Ami-Machi, Ibaraki 300-0335 Japan. Email: yutamn@yahoo.co.jp

McCreight, J.D. USDA-ARS, 1636 E. Alisal St., Salinas CA 93905 USA. Email: jim.mccreight@ars.usda.gov Interests: Melon breeding and genetics

Myers, James R. Dept. Horticulture, Oregon State University, 4037 Ag Life Sciences Bldg., Corvalis OR 97331-7304 USA. Email: myersja@hort.oregonstate.edu

Neill, Amanda. The Botanical Research Inst. of Texas, 509 Pecan St., Fort Worth TX 76102-4060 USA. Email: aneill@brit.org Interests: Gurania and Psiguria

Ng, Timothy J. Dept. Natural Resource Sci., Univ. of Maryland, College Park MD 20742-4452 USA. Email: binkley@umd.edu ; cucurbit.genetics.cooperative@gmail.com Interests: Melon breeding and genetics; postharvest physiology; seed germination

Om, Young-Hyun. #568-3 Pajang-Dong, Jangan -Gu, Suwon 440-853 Republic of Korea. Email: omyh2673@hanmail.net Interests: Breeding of cucurbit vegetables

Owens, Ken. Magnum Seeds, Inc., 5825 Sievers Road, Dixon CA 95620 USA. Email: ken.owens @magnumseeds.com Interests: Cucumber breeding

Ouyang, Wei. Magnum Seeds, 5825 Sievers Rd, Dixon CA 95620 USA. Email: weiouyang1@yahoo.com Interests: Squash, watermelon, cucumber breeding

Palomares, Gloria. Dept. Biotecnología, Univ. Politécnica, Camino de Vera, s/nE-46022 Valencia Spain. Email: gpaloma@btc.upv.es Interests: Genetic improvement in horticultural plants

Paris, Harry. Dept. Vegetable Crops, A.R.O. Newe Ya'ar Research Ctr., P.O. Box 1021, Ramat Yishay 30-095 Israel. Email: hsparis@agri.gov.il Interests: Breeding and genetics of squash and pumpkin

Park, Soon O. TexaAgriLife Res. & Extension Ctre., Texas A & M University, 2415 East Highway 83, Weslaco TX 78596 USA. Email: so-park@tamu.edu Interests: melon genetics

Peiro Abril, José Luis. Apartado de Correos no. 2, E 04720 Aguadulce Spain. Email: peiroab@larural.es; jlp@ramiroarnedo.com Interests: Melon, cantaloupe, watermelon, squash, cucumber breeding, gentics, in vitro

Polewczak, Lisa. Syngenta Seeds, 10290 Greenway Road, Naples FL 34114 USA. Email: l.polewczak @gmail.com Interests: Squash, cantaloupe, watermelon breeding & genetics

Poulos, Jean M. Nunhems USA, Inc., 7087 E. Peltier Rd., Acampo CA 95220 USA. Email: jean.poulos @nunhems.com Interests: Melon breeding

Prabu, Thachinamoorthy. Nunhems India Pvt. Ltd; No 16, Sri Ramanjaneya Complex; 1st 'A' Main, Canara Bank Road; Yelahanka, New Town, Bangalore 64; INDIA; t.prabu@bayer.com; Watermelon breeding

Randhawa, Lakhwinder. Sakata Seed America, Inc., 2854 Niagara Ave, Colusa CA 95932 USA. Email: lrandhawa@sakata.com Interests: Molecular markers

Randhawa, Parm. CA Seed & Plant Lab, 7877 Pleasant Grove Rd., Elverta CA 95626 USA. Email: randhawa @calspl.com

Ray, Dennis. Dept. Plant Sci., Univ. of Arizona, P.O. Box 210036, Tucson AZ 85721-0036 USA. Email: dtray@email.arizona.edu Interests: Genetics and cytogenetics of Cucumis melo and Citrullus spp.

Reitsma, Kathy. North Central Regional Plant Introduction Sta., Iowa State University, Ames IA 50011-1170 USA. Email: kathleen.reitsma@ars.usda.gov Interests: curator of cucurbit germplasm

Reuling, Gerhard T.M. Nunhems Netherlands B.V., P.O. Box 4005, 6080 AA Haelen Limburg Netherlands. Email: g.reuling@nunhems.com Interests: Breeding long cucumber

Robinson, R. W. Emeritus Prof. , Dept. Hort. Sci., NY Agri. Expt. Station, Cornell University, Geneva NY 14456-0462 USA. Email: rwr1@cornell.edu Interests: Breeding and genetics of cucurbits

Rodenburg, Marinus. East-West Seed Indonesia, Desabenteng, Campaka, P.O. Box 1 Purwakarta Indonesia. Email: rien@ewsi.co.id

Rokhman, Fatkhu. PT East West Seed Indonesia, P.O. Box 1, Campaka, Purwakarta 41181, W. Java Indonesia. Email: fatkhu_Rokhman@ewsi.co.id Interests: Cucumber, watermelon and melon breeding

Sheng, Yunyan. Heilongjiang Bayi Agri. Univ.; No. 2 XinFeng Road; High Tech. Development Dist.; Daqing, Heilongjiang 163319; P.R. of CHINA; shengyunyan12345@163.com

Shetty, Nischit V. Monsanto Vegetable Seeds, 2221 CR 832, Felda FL 33930 USA. Email: nischit.shetty @monsanto.com Interests: Cucumber breeding

Shimamoto, Ikuhiro. 146-11 Daigo, Kashihara, NARA 634-0072 Japan. Email: shimamoto@suika-net.co.jp

Simon, Phillip W. USDA-ARS-Vegetable Crops, Dept. of Horticulture, Univ. of Wis., 1575 Linden Dr., Madison WI 53706-1590 USA. Email: psimon@wisc.edu Interests: Breeding and genetics

Stephenson, Andrew G. 208 Mueller Lab, Penn State Univ., University Park PA 16802-5301 USA. Email: as4@psu.edu

Suddath, Pete. Abbot & Cobb, 4429 Mathis Mill Road, Valdosta GA 31602 USA. Email: Pete_suddarth @abbottcobb.com

Sun, Zhanyong. East-West International, Ltd., Rm 2412, New City Int'l Mansion, No 92-1 Minzu AveNanning Guangxi 530021 P.R. China. Email: Zhanyong.sun @eastwestseed.com Interests: Cucurbit breeding

Swanepoel, Cobus. Pannar, P.O. Box 13339, Northmead 1511 South Africa. Email: cobus.swanepoel @pannar.co.za

Tatlioglu, Turan. Martar Tohumculuk A.S., Han Mahallesi, 5 Eylül Caddesi No. 7, Susurluk (Balikesir) 10600 Turkey. Email: turantatlioglu@yahoo.com Interests: Hybrid breeding, sex inheritanc, sex genes

Taurick, Gary. Harris Moran Seed Co., 5820 Research Way, Immokalee FL 34142 USA. Email: g.taurick@hmclause.com Interests: Development of commercial hybrids of pickle, slicer and Beit Alpha cucumbers

Theurer, Christoph. GlaxoSmithKline Consumer Healthcare GmbH & Co. KG, Consumer Healthcare Gmbh&Co.KG, Benzstrasse 25D-71083 Herrenberg Germany. Email: Christoph.Theurer@gsk.com

Tolla, Greg. Monsanto Vegetable Seeds, 37437 State Hwy 16, Woodland CA 95695 USA. Email: greg.tolla @monsanto.com Interests: Breeding and genetics

Vadra Halli, Satish. #1/2, Krishna Road, Basavanagudi, Bangalore 560004 Karnataka India. Email: satishvadrahalli@yahoo.com

Vardi, Eyal. Origene Seeds Ltd., POB 699, Rehovot 76100 Israel. Email: eyal@origeneseeds.com

Wang, Xiqing. Hort. Sub-academy, Heilongjiang Acad. Of Agric. Sci.; No. 666 Haping Road; Harbin 150069; P.R. of CHINA; xiqingwang100@163.com; Watermelon breeding

Webb, Susan E. Dept. of Entomology & Nematology, 2213 Bldg. 970, Univ. of Florida, PO Box 110620, Gainesville FL 32611-0620 USA. Email: sewe@ufl.edu Interests: insect pests of cucurbits

Wehner, Todd. Dept. Horticultural Science, Box 7609, North Carolina State Univ., Raleigh NC 95616 USA. Email: todd_wehner@ncsu.edu Interests: Pickling/slicing cucumber, watermelon, luffa gourd, selection, disease resistance, yield, genetics and breeding

Weng, Yiqun. USDA Vegetable Crops Research Unit, University of Wisconsin, 1575 Linden Dr., Madison WI 53706 USA. Email: yiqun.weng@ars.usda.gov Interests:

Wessel-Beaver, Linda. Dept. Crops & Agroenvironmental Sci., Call Box 9000, Univ. of Puerto Rico, Mayaguez PR 00681-9000 USA. Email: lwesselbeaver@yahoo.com; lindawesselbeaver@upr.edu Interests: Pumpkin and squash breeding and genetics; disease and insect resistance; cucurbit evolution and domestication

Winkler, Johanna. Saatzucht Gleisdorf GmbH, Am Tieberhof 33, A-8200 Gleisdorf Austria. Email: winkler.szgleisdorf@utanet.at

Yang, Xingping. Jiangsu Academy of Agric. Sci.; Institute of Vegetable Crops; No. 50 Zhongling Street; Nanjing 210014 Jiangsu; P.R. of CHINA; xingping@jsmail.com.cn

Yangping, Wang. XiYu Seeds; No. 32 Ningbian East Road; Changji, Xinjiang 831100; P.R. of CHINA; xiyuseeds@sina.com; Watermelon and melon breeding

Zhang, Xingping. Syngenta Seeds, 21435 Co. Rd. 98, Woodland CA 95695 USA. Email: xingping.zhang @syngenta.com Interests: Watermelon and melon genetics & breeding

2010-2011 CGC Membership by Country

Argentina
Lopez-Anido, Fernando

Australia
Herrington, Mark

Austria
Lelley, Tamas
Winkler, Johanna

Brazil
Gardingo, Jose Raulindo
Kobori, Romulo Fujito

China, Peoples Republic of
Haizhen, Li
He, Xiaoming
Liu, Wenge
Luan, Feishi
Ma, Qing
Sheng, Yunyan
Sun, Zhanyong
Wang, Xiqing
Yang, Xingping
Yangping, Wang

Czech Republic
Holman, Bohuslav
Lebeda, Aleš

France
Ignart, Frederic
Majde, Mansour

Germany
Theurer, Christoph

India
Prabu, Thachinamoorthy
Vadra Halli, Satish

Indonesia
Rodenburg, Marinus
Rokhman, Fatkhu

Israel
Cohen, Roni
Davidi, Haim
Karchi, Zvi

Katzir, Nurit
Paris, Harry
Vardi, Eyal

Italy
de Groot, Erik
Ficcadenti, Nadia
Gatto, Gianni
Jones-Evans, Elen

Japan
Furuki, Toshi
Ito, Kimio
Matsumoto, Yuichi
Shimamoto, Ikuhiro

Korea, Republic of
Om, Young-Hyun

Malta
Attard, Everaldo

Mexico
Garza-Ortega, Sergio

Netherlands, The
Hertogh, Kees
Hofstede, Rene
Hoogland, Jan
Kelfkens, Marcel
Reuling, Gerhard T.M.

New Zealand
Grant, Doug

South Africa
Swanepoel, Cobus

Spain
Gil-Albert, Carlos
Palomares, Gloria
Peiro Abril, José Luis

Sweden
Lehmann, Louis Carl

Taiwan
Hu, Cheng-Jung

Thailand
de Hoop, Simon Jan

Turkey
Tatlioglu, Turan

USA
Boyhan, George E.
Bell, Duane
Call, Adam D.
Chung, Paul
Crosby, Kevin
Davis, Angela
Dombrowski, Cory
Everts, Kathryne
Froese, David C.
Goldman, Amy P.
Grumet, Rebecca
Guner, Nihat
Harvey, Michael J.
Jahn, Laboratory of Molly
Johnson, Bill
Johnston, Jr., Robert
Juarez, Benito
Kabelka, Eileen
King, Stephen R.
Kirkbride, Jr., Joseph H.
Knerr, Larry D.
Kole, Chittaranjan
Kumar, Rakesh
Lanini, Brenda
Levi, Amnon
Loy, J. Brent
McCreight, J.D.
Myers, James R.
Neill, Amanda
Ng, Timothy J.
Owens, Ken
Ouyang, Wei
Park, Soon O.
Polewczak, Lisa
Poulos, Jean M.
Randhawa, Lakhwinder
Randhawa, Parm
Ray, Dennis
Reitsma, Kathy
Robinson, R.W.
Shetty, Nischit V.
Simon, Phillip W.

Dane, Fenny
Davis, Angela
Dombrowski, Cory
Everts, Kathryne
Frobish, Mark
Froese, David C.
Gabor, Brad
Goldman, Amy P.
Groff, David
Grumet, Rebecca
Guner, Nihat
Gusmini, Gabriele
Havey, Michael J.
Himmel, Phyllis
Huan, Jin
Jahn, Molly
Johnston, Rob
Juarez, Benito

Kabelka, Eileen
King, Stephen R.
Kirkbride, Jr., Joseph H.
Knerr, Larry D.
Kousik, Chandrasekar (Shaker)
Kumar, Rakesh
Lanini, Brenda
Lee, Chiwon W.
Lester, Gene
Ling, Kai-shu
Lower, Richard L.
Loy, J. Brent
Maynard, Donald N.
McCreight, J.D.
Myers, James R.
Neill, Amanda
Ng, Timothy J.
Ouyang, Wei
Owens, Ken

Park, Soon O.
Polewczak, Lisa
Poulos, Jean M.
Randhawa, Lakhwinder
Randhawa, Parm
Ray, Dennis
Reitsma, Kathy
Robinson, R. W.
Shetty, Nischit V.
Simon, Phillip W.
Stephenson, Andrew G.
Thro, Ann Marie
Tolla, Greg
Wehner, Todd
Weng, Yiqun
Wessel-Beaver, Linda
Williams, Tom V.
Yorty, Paul
Zhang, Xingping

2008-2009 United States CGC Membership By State

Arizona
Ray, Dennis

California
Chung, Paul
Johnson, Bill
Juarez, Benito
Kabelka, Eileen
Kneer, Larry D.
Lanini, Brenda
McCreight, J.D.
Owens, Ken
Ouyang, Wei
Poulos, Jean M.
Randhawa, Lakhwinder
Randhawa, Parm
Tolla, Greg
Zhang, Xingping

Colorado
Froese, David C.

District of Columbia
Kirkbride, Jr., Joseph H.

Florida
Dombrowski, Cory
Guner, Nihat
Polewczak, Lisa
Shetty, Nischit V.
Taurick, Gary
Webb, Susan E.

Georgia
Boyhan, George E.
Suddath, Pete

Iowa
Reitsma, Kathy

Maine
Johnston, Jr., Robert

Maryland

Everts, Kathryne
Ng, Timothy J.

Michigan
Grumet, Rebecca

New Hampshire
Loy, J. Brent

New York
Goldman, Amy P.
Jahn, Laboratory of Molly
Robinson, R.W.

North Carolina
Call, Adam D.
Kumar, Rakesh
Wehner, Todd

Ohio
Bell, Duane

Oklahoma
Davis, Angela

Oregon
Myers, James R.

Pennsylvania
Stephenson, Andrew G.

Puerto Rico
Wessel-Beaver, Linda

South Carolina
Kole, Chittaranjan
Levi, Amnon

Texas
Crosby, Kevin
King, Stephen R.
Neill, Amanda
Park, Soon O.

Wisconsin
Havey, Michael J.
Simon, Phillip W.
Weng, Yiqun

Covenant and By-Laws of the Cucurbit Genetics Cooperative

ARTICLE I. Organization and Purposes

The Cucurbit Genetics Cooperative is an informal, unincorporated scientific society (hereinafter designated "CGC") organized without capital stock and intended not for business or profit but for the advancement of science and education in the field of genetics of cucurbits (Family: Cucurbitaceae). Its purposes include the following: to serve as a clearing house for scientists of the world interested in the genetics and breeding of cucurbits, to serve as a medium of exchange for information and materials of mutual interest, to assist in the publication of studies in the aforementioned field, and to accept and administer funds for the purposes indicated.

ARTICLE II. Membership and Dues

1. The membership of the CGC shall consist solely of active members; an active member is defined as any person who is actively interested in genetics and breeding of cucurbits and who pays biennial dues. Memberships are arranged by correspondence with the Chairman of the Coordinating Committee.
2. The amount of biennial dues shall be proposed by the Coordinating Committee and fixed, subject to approval at the Annual Meeting of the CGC. The amount of biennial dues shall remain constant until such time that the Coordinating Committee estimates that a change is necessary in order to compensate for a fund balance deemed excessive or inadequate to meet costs of the CGC.
3. Members who fail to pay their current biennial dues within the first six months of the biennium are dropped from active membership. Such members may be reinstated upon payment of the respective dues.

ARTICLE III. Committees

1. The Coordinating Committee shall govern policies and activities of the CGC. It shall consist of six members elected in order to represent areas of interest and importance in the field. The Coordinating Committee shall select its Chairman, who shall serve as a spokesman of the CGC, as well as its Secretary and Treasurer.
2. The Gene List Committee, consisting of at least five members, shall be responsible for formulating rules regulating the naming and symbolizing of genes, chro-

mosomal alterations, or other hereditary modifications of the cucurbits. It shall record all newly reported mutations and periodically report lists of them in the Report of the CGC. It shall keep a record of all information pertaining to cucurbit linkages and periodically issue revised linkage maps in the Report of the CGC. Each committee member shall be responsible for genes and linkages of one of the following groups: cucumber, *Cucurbita* spp., muskmelon, watermelon, and other genera and species.
3. Other committees may be selected by the Coordinating Committee as the need for fulfilling other functions arises.

ARTICLE IV. Election and Appointment of Committees

1. The Chairman will serve an indefinite term while other members of the Coordinating Committee shall be elected for ten-year terms, replacement of a single retiring member taking place every other year. Election of a new member shall take place as follows: A Nominating Committee of three members shall be appointed by the Coordinating Committee. The aforesaid Nominating Committee shall nominate candidates for an anticipated opening on the Coordinating Committee, the number of nominees being at their discretion. The nominations shall be announced and election held by open ballot at the Annual Meeting of the CGC. The nominee receiving the highest number of votes shall be declared elected. The newly elected member shall take office immediately.
2. In the event of death or retirement of a member of the Coordinating Committee before the expiration of his/her term, he/she shall be replaced by an appointee of the Coordinating Committee.
3. Members of other committees shall be appointed by the Coordinating Committee.

ARTICLE V. Publications

1. One of the primary functions of the CGC shall be to issue an Annual Report each year. The Annual Report shall contain sections in which research results and information concerning the exchange of stocks can be published. It shall also contain the annual financial statement. Revised membership lists and

other useful information shall be issued periodically. The Editor shall be appointed by the Coordinating Committee and shall retain office for as many years as the Coordinating Committee deems appropriate.

2. Payment of biennial dues shall entitle each member to a copy of the Annual Report, newsletters, and any other duplicated information intended for distribution to the membership. The aforementioned publications shall not be sent to members who are in arrears in the payment of dues. Back numbers of the Annual Report, available for at least the most recent five years, shall be sold to active members at a rate determined by the Coordinating Committee.

ARTICLE VI. Meetings

An Annual Meeting shall be held at such time and place as determined by the Coordinating Committee. Members shall be notified of time and place of meetings by notices in the Annual Report or by notices mailed not less than one month prior to the meeting. A financial report and information on enrollment of members shall be presented at the Annual Meeting. Other business of the Annual Meeting may include topics of agenda selected by the Coordinating Committee or any items that members may wish to present.

ARTICLE VII. Fiscal Year

The fiscal year of the CGC shall end on December 31.

ARTICLE VIII. Amendments

These By-Laws may be amended by simple majority of members voting by mail ballot, provided a copy of the proposed amendments has been mailed to all the active members of the CGC at least one month previous to the balloting deadline.

ARTICLE IX. General Prohibitions

Notwithstanding any provisions of the By-Laws or any document that might be susceptible to a contrary interpretation:

1. The CGC shall be organized and operated exclusively for scientific and educational purposes.

2. No part of the net earnings of the CGC shall or may under any circumstances inure to the benefit of any individual.

3. No part of the activities of the CGC shall consist of carrying on propaganda or otherwise attempting to influence legislation of any political unit.

4. The CGC shall not participate in, or intervene in (including the publishing or distribution of statements), any political campaign on behalf of a candidate for public office.

5. The CGC shall not be organized or operated for profit.

6. The CGC shall not:
   a. lend any part of its income or corpus without the receipt of adequate security and a reasonable rate of interest to;
   b. pay any compensation in excess of a reasonable allowance for salaries or other compensation for personal services rendered to;
   c. make any part of its services available on a preferential basis to;
   d. make any purchase of securities or any other property, for more than adequate consideration in money's worth from;
   e. sell any securities or other property for less than adequate consideration in money or money's worth; or
   f. engage in any other transactions which result in a substantial diversion of income or corpus to any officer, member of the Coordinating Committee, or substantial contributor to the CGC.

The prohibitions contained in this subsection (6) do not mean to imply that the CGC may make such loans, payments, sales, or purchases to anyone else, unless authority be given or implied by other provisions of the By-Laws.

ARTICLE X. Distribution on Dissolution

Upon dissolution of the CGC, the Coordinating Committee shall distribute the assets and accrued income to one or more scientific organizations as determined by the Committee, but which organization or organizations shall meet the limitations prescribed in sections 1-6 of Article IX.

Gene Nomenclature for the Cucurbitaceae

1. Names of genes should describe a characteristic feature of the mutant type in a minimum of adjectives and/or nouns in English or Latin.

2. Genes are symbolized by italicized Roman letters, the first letter of the symbol being the same as that for the name. A minimum number of additional letters are added to distinguish each symbol.

3. The first letter of the symbol and name is capitalized if the mutant gene is dominant,. All letters of the symbol and name are in lower case if the mutant gene is recessive, with the first letter of the symbol capitalized for the dominant or normal allele. (Note: For CGC *research articles*, the normal allele of a mutant gene may be represented by the symbol "+", or the symbol of the mutant gene followed by the superscript "+", if greater clarity is achieved for the manuscript.)

4. A gene symbol shall not be assigned to a character unless supported by statistically valid segregation data for the gene.

5. Mimics, i.e. different mutants having similar phenotypes, may either have distinctive names and symbols or be assigned the same gene symbol, followed by a hyphen and distinguishing Arabic numeral or Roman letter printed at the same level as the symbol. The suffix "-1" is used, or may be understood and not used, for the original gene in a mimic series. It is recommended that allelism tests be made with a mimic before a new gene symbol is assigned to it.

6. Multiple alleles have the same symbol, followed by a Roman letter or Arabic number superscript. Similarities in phenotype are insufficient to establish multiple alleles; the allelism test must be made.

7. Indistinguishable alleles, i.e. alleles at the same locus with identical phenotypes, preferably should be given the same symbol. If distinctive symbols are assigned to alleles that are apparent re-occurrences of the same mutation, however, they shall have the same symbol with distinguishing numbers or letters in parentheses as superscripts.

8. Modifying genes may have a symbol for an appropriate name, such as intensifier, suppressor, or inhibitor, followed by a hyphen and the symbol of the allele affected. Alternatively, they may be given a distinctive name unaccompanied by the symbol of the gene modified.

9. In cases of the same symbol being assigned to different genes, or more than one symbol designated for the same gene, priority in publication will be the primary criterion for establishing the preferred symbol. Incorrectly assigned symbols will be enclosed in parentheses on the gene lists.

10. The same symbol shall not be used for nonallelic genes of different *Cucurbita* species. Allelic genes of compatible species are designated with the same symbol for the locus.

References:

1. CGC Gene List Committee. 1982. Update of cucurbit gene list and nomenclature rules. CGC 5:62-66.

2. Rieger, R., A. Michaelis and M.M. Green. 1976. Glossary of Genetics and Cytogenetics (4th ed.). Springer-Verlag.

3. Robinson, R.W., H.M. Munger, T.W. Whitaker and G.W. Bohn. 1976. Genes of the Cucurbitaceae. HortScience 11:554-568.